BIOTECHNOLOGY AND RENEWABLE ENERGY

BIOTECHNOLOGY
AND RENEWABLE ENERGY

Edited by

MURRAY MOO-YOUNG
Faculty of Engineering, University of Waterloo, Ontario, Canada

SADIQ HASNAIN
National Research Council of Canada, Ottawa, Ontario, Canada

and

JONATHAN LAMPTEY
Pioneer Hi-Bred International Inc., Johnston, Iowa, USA

ELSEVIER APPLIED SCIENCE PUBLISHERS
LONDON and NEW YORK

ELSEVIER APPLIED SCIENCE PUBLISHERS LTD
Crown House, Linton Road, Barking, Essex IG11 8JU, England

Sole Distributor in the USA and Canada
ELSEVIER SCIENCE PUBLISHING CO., INC.
52 Vanderbilt Avenue, New York, NY 10017, USA

WITH 64 TABLES AND 87 ILLUSTRATIONS

© ELSEVIER APPLIED SCIENCE PUBLISHERS LTD 1986

British Library Cataloguing in Publication Data

Biotechnology and renewable energy.
1. Biomass energy
I. Moo-Young, Murray   II. Hasnain, Sadiq
III. Lamptey, Jonathan
662'.6   TP360

Library of Congress Cataloging in Publication Data

Biotechnology and renewable energy.

Bibliography: p.
Includes index.
1. Biomass energy.   2. Biomass chemicals.
I. Moo-Young, M.   II. Hasnain, Sadiq.   III. Lamptey,
Jonathan.
TP360.B5958   1986   662'.8   86-19805

ISBN 1-85166-061-5

Printed in Great Britain by Galliard (Printers) Ltd, Great Yarmouth

PREFACE

As petroleum reserves decline, global attention is turned to cellulosic biomass as a possible source of renewable energy. Inevitably, biotechnology becomes part of the scenario. In this volume, experts primarily from the USA and Canada explore aspects of biotechnology as they are involved in the production and use of biomass feedstocks in the manufacture of a variety of end products and intermediate products. The material should be of interest to those involved in lignocellulose biotechnologies in general and bioenergy in particular. Students, researchers, industrialists and government planners are addressed.

The material is treated under the following seven Section headings:

Section 1: Biomass Production
Section 2: Pretreatment Technology
Section 3: Enzyme Production
Section 4: Genetic Engineering Research
Section 5: Fuels and Chemicals
Section 6: Fermentation Technology
Section 7: Process Implementation

The manuscripts have not been subjected to an external referee system in the sense that only the authors with the approval of the editors have decided on the technical content of the published material. We are also grateful to the National Research Council of Canada for providing financial support in the preparation of the printed material.

We thank Mrs Arlene Lamptey for her dedicated assistance in editorial structuring. Without her, this effort would have been in vain.

MURRAY MOO-YOUNG
Waterloo, Ontario
SADIQ HASNAIN
Ottawa, Ontario
JONATHAN LAMPTEY
Johnston, Iowa

v

CONTENTS

Preface v

List of Contributors xi

Section 1: Biomass Production

Micropropagation of Conifers in New Zealand: Radiata Pine a
Case History 1
T. A. THORPE, J. AITKEN-CHRISTIE, K. HORGAN and D. R. SMITH

Shoot Primordia Induction on Bud Explants from 12–15 Year Old
Douglas Fir: The Requirement for BA and NH_4NO_3 in
Dormant Collections 12
D. I. DUNSTAN, G. H. MOHAMMED and T. A. THORPE

In vitro Propagation of Mature Conifers 20
J. M. BONGA

Biotechnology in Breeding for Biomass Production . . . 23
R. H. HO and L. ZSUFFA

Section 2: Pretreatment Technology

Steam Pretreatment of Aspenwood for Enhanced Enzymatic
Hydrolysis 36
H. H. BROWNELL, M. MES-HARTREE and J. N. SADDLER

Pretreatment of Lignocellulosics for Bioconversion Applications:
Process Options 46
J. LAMPTEY, M. MOO-YOUNG and C. W. ROBINSON

Section 3: Enzyme Production

A New Approach in Solid State Fermentation for Cellulase
Production 57
D. S. CHAHAL

Production and Use of Cellulases in the Conversion of Cellulose to
Fuels and Chemicals 70
A. W. KHAN

Bacterial Cellulases 76
C. R. MACKENZIE

Factors Affecting Cellulase Production and the Efficiency of
Cellulose Hydrolysis 83
J. N. SADDLER, C. M. HOGAN, G. LOUIS-SEIZE and E. K. C. YU

Cellulase from an Acidophilic Fungus. 93
A. LEDUY

Section 4: Genetic Engineering Research

The Characteristics and Cloning of Bacterial Cellulases . . 101
C. W. FORSBERG, K. TAYLOR, B. CROSBY and D. Y. THOMAS

In Search of Lignin Modifying Genes. 112
P. CHARTRAND, P. TRUDEL, D. COURCHESNE and C. ROY

Genetic Engineering of Cellulase Genes from a Fungus . . 118
V. L. SELIGY, M. J. DOVE, M. YAGUCHI, G. E. WILLICK and
F. MORANELLI

The Development of *Azotobacter* as a Bacterial Fertilizer by the
Introduction of Exogenous Cellulase Genes 125
B. R. GLICK, J. J. PASTERNAK and H. E. BROOKS

Section 5: Fuels and Chemicals

Butanediol Production from Cellulose and Hemicellulose . . 135
E. K. C. YU and J. N. SADLER

Optimization of 2,3 Butanediol Production from Glucose in Batch
and Fed-Batch Cultures of *Bacillus polymyxa* . . . 145
V. M. LAUBE, D. GROLEAU, S. M. MARTIN and I. J. MCDONALD

Biological Production of Economically-Recoverable Products from
Dilute Ethanol Streams. 153
D. W. ARMSTRONG, S. M. MARTIN and H. YAMAZAKI

Recent Progress in Obtaining Ethanol from Xylose . . . 161
H. SCHNEIDER, A. P. JAMES, J. LABELLE, H. LEE, G.
MAHMOURIDES, R. MALESZKA, G. CALLEJA and S. LEVY-RICK

Nutritional Requirements of *Clostridium acetobutylicum* ATCC 824 167
F. W. WELSH and I. A. VELIKY

The Use of Renewable Carbohydrate Sources for the Microbial
Production of Acetone and Butanol 175
C. W. FORSBERG. S. F. LEE, H. SCHELLHORN, L. N. GIBBINS,
E. MASON and F. MAINE

Production of Alcohol by Flocculating Yeast 187
N. KOSARIC, A. WIECZOREK and Z. DUVNJAK

Section 6: Fermentation Technology

The Development of Variable Surface Reactors for Bioconversions 200
P. P. MATTEAU and C. E. CHOMA

Kinetic Modelling and Computer Control of Continuous Alcoholic
Fermentation in the Gas-Lift Tower Fermenter . . . 208
D. M. COMBERBACH, C. GHOMMIDH and J. D. BU'LOCK

Ethanol Production by Adsorbed Cell Bioreactors . . . 225
A. J. DAUGULIS and D. E. SWAINE

Early Developments of a Novel Fermenter-Purifier . . . 234
D. L. MULHOLLAND

Rapid Production of High Concentrations of Ethanol Using
Unmodified Industrial Yeast 246
W. M. INGLEDEW and G. P. CASEY

Bioprocess Modelling and Optimization using Acetone–Butanol
System 258
B. VOLESKY

High Rate Reactors for Methane Production 268
L. VAN DEN BERG

Section 7: Process Implementation

Scale-up of the Bio-Hol Process for the Conversion of Biomass to
Ethanol 276
G. R. LAWFORD, R. CHARLEY, R. EDAMURA, J. FEIN,
K. HOPKINS, D. POTTS, B. ZAWADZKI and H. LAWFORD

Commercial Update on the Stake Process 286
J. D. TAYLOR

LIST OF CONTRIBUTORS

J. AITKEN-CHRISTIE
    Forest Research Institute, Rotorua, New Zealand

D. W. ARMSTRONG
    Division of Biological Sciences, National Research Council of Canada, Ottawa, Ontario K1A 0R6, Canada

L. VAN DEN BERG
    Division of Biological Sciences, National Research Council of Canada, Ottawa, Ontario K1A 0R6, Canada

J. M. BONGA
    Canadian Forestry Service, Maritimes Forest Research Centre, PO Box 4000, Fredericton, N.B., E3B 5P7, Canada

H. E. BROOKS
    Department of Biology, University of Waterloo, Waterloo, Ontario N2L 3G1, Canada

H. H. BROWNELL
    Forintek Canada Corporation, 800 Montreal Road, Ottawa K1G 3Z5, Canada

J. D. BU'LOCK
    Weizmann Microbial Chemistry Laboratory, University of Manchester, Oxford Road, Manchester M13 9PL, UK

G. CALLEJA
    Ethanol from Cellulose Program, Canada

xi

G. P. CASEY
Department of Physiology, Carlsberg Laboratory, GL Carlsberg VEJ 10, DK 2500, Copenhagen, Valby, Denmark

D. S. CHAHAL
Bacteriology Research Centre, Institut Armand-Frappier, University of Québec, 531 Boulevard des Prairies, Laval, Québec H7V 1B7, Canada

R. CHARLEY
Weston Research Centre, 1047 Yonge Street, Toronto, Ontario M4W 2L3, Canada

P. CHARTRAND
Department of Microbiology, Medical Faculty, University of Sherbrooke, Sherbrooke, Québec J1K 2R1, Canada

C. E. CHOMA
Queen's University, Kingston, Ontario K7L 3N6, Canada

D. M. COMBERBACH
Allelix Inc., 6850 Goreway Drive, Mississauga, Ontario L4V 1P1, Canada

D. COURCHESNE
Department of Microbiology, Medical Faculty, University of Sherbrooke, Sherbrooke, Québec J1K 2R1, Canada

B. CROSBY
Plant Biotechnology Institute, National Research Council of Canada, Saskatoon, Saskatchewan S7N 0W9, Canada

A. J. DAUGULIS
Department of Chemical Engineering, Queen's University, Kingston, Ontario K7L 3N6, Canada

M. J. DOVE
Molecular Genetics Section, Division of Biological Sciences, National Research Council of Canada, Ottawa, Ontario K1A 0R6, Canada

D. I. DUNSTAN
Plant Tissue Culture Research Group, Kelowna Nurseries Ltd, PO Box 178, Kelowna, B.C., V1Y 7N5, Canada

Z. DUVNJAK
Chemical and Biochemical Engineering, Faculty of Engineering Science, University of Western Ontario, London, Ontario N6A 5B9, Canada

R. EDAMURA
Weston Research Centre, 1047 Yonge Street, Toronto, Ontario M4W 2L3, Canada

J. FEIN
Weston Research Centre, 1047 Yonge Street, Toronto, Ontario M4W 2L3, Canada

C. W. FORSBERG
Department of Microbiology, University of Guelph, Guelph, Ontario N1G 2W1, Canada

C. GHOMMIDH
Laboratoire de Génie Microbiologique, Université des Sciences et Techniques du Languédoc, Place E Bataillon, 34060 Montpellier Cédex, France

L. N. GIBBINS
Department of Microbiology, University of Guelph, Guelph, Ontario N1G 2W1, Canada

B. R. GLICK
Department of Biology, University of Waterloo, Waterloo, Ontario N2L 3G1, Canada

D. GROLEAU
Division of Biological Sciences, National Research Council of Canada, Ottawa, Ontario K1A 0R6, Canada

R. H. HO
Ontario Tree Improvement and Forest Biomass Institute, Ministry of Natural Resources, Maple, Ontario, Canada

xiv

C. M. Hogan
Biotechnology and Chemistry Department, Forintek Canada Corporation, 800 Montreal Road, Ottawa, Ontario K1G 3Z5, Canada.

K. Hopkins
Weston Research Centre, 1047 Yonge Street, Toronto, Ontario M4W 2L3, Canada

K. Horgan
Forest Research Institute, Rotorua, New Zealand

W. M. Ingledew
Applied Microbiology and Food Science, University of Saskatchewan, Saskatoon, Saskatchewan S7N 0W0, Canada

A. P. James
National Research Council of Canada, Ottawa, Ontario K1A 0R6, Canada

A. W. Khan
Division of Biological Sciences, National Research Council of Canada, Ottawa, Ontario K1A 0R6, Canada

N. Kosaric
Chemical and Biochemical Engineering, Faculty of Engineering Science, University of Western Ontario, London, Ontario N6A 5B9, Canada

J. Labelle
National Research Council of Canada, Ottawa, Ontario K1A 0R6, Canada

J. Lamptey
Process Biotechnology Research Group, Department of Chemical Engineering, University of Waterloo, Waterloo, Ontario N2L 3G1, Canada

V. M. Laube
Division of Biological Sciences, National Research Council of Canada, Ottawa, Ontario K1A 0R6, Canada

G. R. LAWFORD
Weston Research Centre, 1047 Yonge Street, Toronto, Ontario M4W 2L3, Canada

H. LAWFORD
Department of Biochemistry, University of Toronto, Toronto, Ontario M5S 1A8, Canada

D. LEDUY
Department of Chemical Engineering, Laval University, Sainte-Foy, Québec G1K 7P4, Canada

H. LEE
National Research Council of Canada, Ottawa, Ontario K1A 0R6, Canada

S. F. LEE
Department of Microbiology, University of Guelph, Guelph, Ontario N1G 2W1, Canada

S. LEVY-RICK
Ethanol from Cellulose Program, Canada

G. LOUIS-SEIZE
Biotechnology of Chemistry Department, Forintek Canada Corporation, 800 Montreal Road, Ottawa, Ontario K1G 3Z5, Canada

I. J. MCDONALD
Division of Biological Sciences, National Research Council of Canada, Ottawa, Ontario K1A 0R6, Canada

C. R. MACKENZIE
Division of Biological Sciences, National Research Council of Canada, Ottawa, Ontario K1A 0R6, Canada

G. MAHMOURIDES
National Research Council of Canada, Ottawa, Ontario K1A 0R6, Canada

F. MAINE
Frank Maine Consulting, 71 Sherwood Drive, Guelph, Ontario N1E 6E6, Canada

R. MALESZKA
National Research Council of Canada, Ottawa, Ontario K1A 0R6, Canada

S. M. MARTIN
Division of Biological Sciences, National Research Council of Canada, Ottawa, Ontario K1A 0R6, Canada

E. MASON
Frank Maine Consulting, 71 Sherwood Drive, Guelph, Ontario N1E 6E6, Canada

P. P. MATTEAU
National Research Council of Canada, Bioenergy Program, Ottawa, Ontario K1A 0R6, Canada

M. MES-HARTREE
Forintek Canada Corporation, 800 Montreal Road, Ottawa, Ontario K1G 3Z5, Canada

G. H. MOHAMMED
Plant Tissue Culture Research Group, Kelowna Nurseries Ltd, PO Box 178, Kelowna, B.C., V1Y 7N5, Canada

M. YOO-YOUNG
Process Biotechnology Research Group, Department of Chemical Engineering, University of Waterloo, Waterloo, Ontario N2L 3G1, Canada

F. MORANELLI
Molecular Genetics Section, Division of Biological Sciences, National Research Council of Canada, Ottawa, Ontario K1A 0R6, Canada

D. L. MULHOLLAND
Ontario Research Foundation, Canada

J. J. PASTERNAK
Department of Biology, University of Waterloo, Waterloo, Ontario N2L 3G1, Canada

D. POTTS
Weston Research Centre, 1047 Yonge Street, Toronto, Ontario M4W 2L3, Canada

C. W. ROBINSON
Process Biotechnology Research Group, Department of Chemical Engineering, University of Waterloo, Waterloo, Ontario N2L 3G1, Canada

C. ROY
Department of Chemical Engineering, Applied Science Faculty, University of Sherbrooke, Sherbrooke, Quebec J1K 2R1, Canada

J. N. SADDLER
Biotechnology and Chemistry Department, Forintek Canada Corporation, 800 Montreal Road, Ottawa, Ontario K1G 3Z5, Canada

H. SCHELLHORN
Department of Microbiology, University of Guelph, Guelph, Ontario N1G 2W1, Canada

H. SCHNEIDER
National Research Council of Canada, Ottawa, Ontario K1A 0R6, Canada

V. L. SELIGY
Molecular Genetics Section, Division of Biological Sciences, National Research Council of Canada, Ottawa, Ontario K1A 0R6, Canada

D. R. SMITH
Forest Research Institute, Rotorua, New Zealand

D. W. SWAINE
Department of Chemical Engineering, Queen's University, Kingston, Ontario K7L 3N6, Canada

xviii

J. D. TAYLOR
Stake Technology Limited, 208 Wyecroft Road, Oakville, Ontario L6K 3T8, Canada

K. TAYLOR
Department of Microbiology, University of Guelph, Guelph, Ontario N1G 2W1, Canada

D. Y. THOMAS
Biotechnology Research Institute, National Research Council of Canada, Ottawa, Ontario K1A 0R6, Canada

T. A. THORPE
Plant Physiology Research Group, Department of Biology, University of Calgary, Calgary, Alberta T2N 1N4, Canada

P. TRUDEL
Department of Microbiology, Medical Faculty, University of Sherbrooke, Sherbrooke, Québec, J1K 2R1, Canada

I. A. VELIKY
Division of Biological Sciences, National Research Council of Canada, Ottawa, Ontario K1A 0R6, Canada

B. VOLESKY
Biochemical Engineering Unit, McGill University, Montreal, Québec H3A 2A7, Canada

F. W. WELSH
Division of Biological Sciences, National Research Council of Canada, Ottawa, Ontario K1A 0R6, Canada

A. WIECZOREK
Institute of Chemical Engineering, Technical University of Lodz, 90–924 Lodz, ul Wolczanska 175, Poland

G. E. WILLICK
Molecular Genetics Section, Division of Biological Sciences, National Research Council of Canada, Ottawa, Ontario K1A 0R6, Canada

xix

M. YAGUCHI
Molecular Genetics Section, Division of Biological Sciences, National Research Council of Canada, Ottawa, Ontario K1A 0R6, Canada

H. YAMAZAKI
Department of Biology, Carleton University, Ottawa, Ontario K1S 5B6, Canada

E. K. C. YU
Biotechnology and Chemistry Department, Forintek Canada Corporation, 800 Montreal Road, Ottawa, Ontario K1G 3Z5, Canada

B. ZAWADZKI
Weston Research Centre, 1047 Yonge Street, Toronto, Ontario M4W 2L3, Canada

L. ZSUFFA
Faculty of Forestry, University of Toronto, Toronto, Ontario M5S 1A1, Canada

MICROPROPAGATION OF CONIFERS IN NEW ZEALAND: RADIATA PINE A CASE HISTORY

Trevor A. Thorpe[1], Jenny Aitken-Christie[2],
Kathryn Horgan[2] and Dale R. Smith[2]

[1]Plant Physiology Research Group, Department of Biology,
University of Calgary, Calgary, Alberta T2N 1N4

[2]Forest Research Institute, Rotorua, New Zealand

ABSTRACT

Radiata pine (Pinus radiata D. Don) is the most important forest tree species in New Zealand and represents 80% of the annual timber cut in sawmills. It has been subjected to tree improvement and intensive management trials for over 30 years. Vegetative propagation trials also began early for clonal orchard development and for afforestation. Grafting and rooted cuttings are now being supplemented with tissue culture methods. Micropropagation studies with excised mature embryos began in the 1970s and in 1979 a small production laboratory was set up. The present methods used involve several distinct steps, namely shoot initiation on excised embryos or cotyledons, elongation, root initiation and development, and plantlet hardening. These methods produce about 250 shoots/clone and over 80% rooting. Work is also taking place with mature explants. Studies on cold storage of developed shoots and remultiplication are also underway, mainly as a tool for facilitating clonal testing of material. Labour is the main component in the cost of micropropagated material, which is ca NZ $450 per thousand plantlets. Several uses of micropropagated radiata pine are now envisaged by both government and industry in New Zealand. Present research activity is geared towards cost reduction and extension of the usefulness of the technology in the short term for mass vegetative propagation and in the longer term for clonal forestry in the stricter sense.

INTRODUCTION

In New Zealand, a country with a temperate maritime climate, the forests make up 27% of the total land area and cover 7.1×10^6 hectares, of which 9×10^5 ha are man-made (Wilcox 1983). Yearly about 4×10^4 ha of land of low agricultural productivity are converted to forests and another 1.6×10^4 ha are replanted following the logging of mature plantations. These require 7×10^7 trees annually. In 1981, 9.1×10^3 m^3 of wood was harvested and about half was exported mainly to Australia and Japan. The major items exported, were newsprint, kraft pulp, timber, fibreboard and plywood, mechanical pulp, kraft papers and paperboard, logs, chips and chemicals (Wilcox, 1983).

Radiata pine (Pinus radiata D. Don), a native of California, USA, is the most successful of all introduced forest tree species in New Zealand. It is thus the foremost species planted and constitutes ca. 90% of the

present area of softwood plantations, an area of 8 X 10^5 ha (Wilcox, 1983).
It grows faster and on a wider range of sites than any other conifer. Pinus
radiata is the species preferred by the forest products industries, and
accounts for almost all the export earnings, representing 80% of the timber
cut in saw mills.

Initial plantings of radiata pine in New Zealand date back over a
century and were for farm shelter belts (Wilcox, 1983). Their rapid growth
led to the establishment of extensive pine forests in the 1920s and 1930s
from seed collected from these early shelter belts. This work was carried
out under MacIntosh-Ellis, a Canadian, who was the first Director of the
Forest Service (Sutton, 1984). Genetic variation within stands became
evident as trees exhibited a tremendous array of branching behaviour and
stem form. Tree improvement programmes with radiata pine have been in
progress for over 30 years. In 1982 over 1200 ha were in various
experiments, progeny tests, clonal archives, seed orchards and provenance
plantations in over 150 separate stands or plots in New Zealand (Wilcox,
1983). This represents a genetic estate of about 5 X 10^5 trees that are
periodically measured, and utilized in selection or propagation. Similarly
silvicultural management based on Ure's principles with continued
modifications and improvements have been undertaken since the early 1950s
(Sutton, 1984). The basic principles of early thinning to waste, followed by
a commercial production thinning in the rotation, now include prunning to
produce defect free clearwood in the lower logs. Most radiata pine seeds
presently used are of genetically improved varieties produced from
progeny-tested parents in clonal seed orchards (Wilcox, 1983). Continuing
selection programmes at the Forest Research Institute (FRI) are aimed at
improving the rate of diameter growth, straightness of the stem, size and
distribution of the branches, and resistance to needle diseases.

Plantation forestry in New Zealand is based on intensive management
(Sutton, 1984). Current silvicultural practice is aimed at harvesting trees
25-30 years old from stands at final densities of 200-300 stems per ha.
Initial planting is between 1500 and 2000 seedlings per ha. Thinnings to
waste are usually made during the first 10 years to give the required crop
density. Thinnings may also be extracted for posts or pulp at around 11-15
years. The objective of such intensive tending is to produce defect-free
wood for veneers or finishing grade timber from pruned logs, framing-grades
from suitable unpruned portions, and reconstituted wood products or pulp
from the remainder (Smith, 1984a).

VEGETATIVE PROPAGATION IN TREE IMPROVEMENT

Tree improvement research involves at least three phases, namely
selection, breeding and progeny testing. Vegetative propagation is an
integral part of any tree improvement programme and can play various roles.
These include holding of genotypes in archives; providing clonal seed
orchards; mass propagation either of select clones for crop uniformity and
full utilization of non-additive gene effects or of scarce or expensive
seedling stock; provision of genetic information; by testing candidate
genotypes for mass clonal propagation, or screening seed parents for
breeding value, or estimating genetic parameters; providing material for
research into physiology and pathology; and for potential applications such
as genetic engineering (Burdon, 1982). At the same time that the tree
breeding programme was being developed in New Zealand, the attractive option

for direct clonal afforestation was also investigated, because of some of the above advantages (Smith, 1984a).

The traditional methods for vegetative propagation are rooted cuttings, or rooted needle fasicles and grafting. While grafting gives the quickest propagation of select parents, graft incompatability is a major limitation. The first round of tree selection in radiata pine, which was undertaken from 1953 to 1958, utilized grafts for setting up seed orchards. These developed imcompatability problems between the ages of 5 and 15 years, causing losses of up to 60% depending on site (Anon, 1979). Subsequent seed orchard establishment has been with rooted cuttings, although grafted stock are still being used as a temporary step to provide cuttings 3 years after the grafted plants have been established (Anon, 1983). A variety of scions have been used (T. Faulds, per. comm). These include scions with or without pollen catkins or needles, and 7 to 9-month-old un-rootpruned seedlings are used as rootstock.

Although techniques have been developed to produce rooted cuttings of radiata pine of any age, extensive field testing has shown that there may be disadvantages in the use of old clones for afforestation. These early studies indicated that seedlings had early superiority in relative height growth rate over cuttings (Sweet and Wells, 1974). The volume loss at the fifth year after planting was calculated to be 23% for cuttings from 10-year-old ortets and 32% for cuttings from 23-year-old ortets compared with seedlings. Although these losses are now believed to have been compounded by poor root form in the cuttings from the older ortets, other studies have also demonstrated an effect of maturation of ortets on growth rate of their cuttings (Shelbourne and Thulin, 1974). Propagules taken from 5 year-old ortets had a slower average growth rate than seedlings. A large variation between clones was observed in most characters, particularly growth rate; but it was apparent that selection of the best clones would lead to substantial gains in most characters over the mean of the clones tested. Furthermore it can be concluded from their data that the best clones performed better than the seedling means (Smith, 1984a). The use of cuttings allows selection of genotypes with straighter stems, free from malformation and with finer branches. The demonstration that some clones performed better than the seedlings mean for growth indicates that a trade-off of form against growth is not an inevitable consequence of the use of cuttings (Smith, 1984b). Recent work indicates that year-old seedlings from controlled pollinated seed of the best crosses can be vegetatively multiplied by a factor of 4 to 5 or higher in the nursery or by a factor of up to 20 from 2 to 4-year-old trees planted in the field using stem cuttings (Anon 1983).

Tasman Forestry Ltd., a subsidiary of Fletcher Challenge Ltd., New Zealand's 3rd largest public company and recent purchaser of Crown Zellerbach (Canada) Ltd., has already embraced the use of cuttings in their afforestation programme. Rooted cuttings from field grown trees are being produced in their forest nursery and are being used in cluster planting at a rate of 750 stems per ha to obtain a final density of 250 trees per ha. Results from production trials using seed orchard progeny and cuttings showed that the average per tree volume was greater in the cuttings than the seedlings at 9 years (Gleed, 1983).

Tissue culture technology represents an additional means of vegetative propagation. The main advantages of this method are its potentials for rapid and large rates of multiplication. Thus while a rooted cutting usually produces a single plant from which some years later further cuttings can be taken, even the most limited tissue culture system, i.e., use of resting buds, can produce several axillary as well as adventitious shoots (Biondi and Thorpe 1982; Thorpe and Biondi (1984). Both of these can form additional axillary and/or adventitious shoots, often within a matter of weeks or at most months. In addition tissue culture approaches open up scores of other possible uses in tree improvement (Thorpe and Biondi 1984).

Since the early 1970s studies on the tissue culture propagation or micropropagation of radiata pine have been undertaken at FRI. These studies led to the demonstration in 1977 that plantlets could be regenerated from mature embryos (Reilly and Washer 1977). Improvements and modifications have increased the efficiency of the process. The development of the present methods being used have been given in several publications (Aitken et al., 1981; Horgan and Aitken, 1981; Horgan, 1982; Aitken-Christie et al., 1982; Aitken-Christie and Thorpe, 1984; Smith, 1984a).

MICROPROPAGATION OF RADIATA PINE

The initial micropropagation methods developed utilized excised embryos and cotyledons from germinated seed as explants. Seeds were sterilized, stratified, resterilized prior to excising the embryo or placing on a sterile germinating medium. Excised viable (white) embryos were laid on the medium lengthwise or inverted so that the cotyledons were in intimate contact with the medium. Alternately seeds were germinated in the dark at 25°C and cotyledons excised after the radical had emerged for 1 day i.e. 5-14 days after, depending on seed source. Plantlet formation involves several distinct steps, namely shoot initiation, elongation, multiplication, root initiation, development and plantlet hardening (see Aitken-Christie and Thorpe, 1984; Smith, 1984a for details). In brief, shoot initiation occurs on a medium consisting of either Schenk and Hildebrandt"s (SH, 1972) or Quoirin and Le Poivre (LP, 1977) salts, vitamins, sucrose (3% w/v) and N^6-benzyladenine (22.2 μM) under 16 hr photoperiod (ave. intensity 80 $\mu Em^{-2}s^{-1}$) and day/night temp 23-28/18-23°C. After 3 weeks the explants are transferred to fresh medium lacking cytokinin. Continued transfer at 4-6 week intervals allows for development and multiplication of the shoots. Clusters of shoots are subdivided at transfer to facilitate both processes. When shoots reach a size of 1.0 to 1.5 cm, they are remultiplied, cold stored (see below) or rooted. Three types of adventitious shoots are produced. These have been termed waxy, wet or translucent. The latter do not survive and are discarded. The wet shoots often have needles which are stuck together and lack epicuticular waxes, whereas the waxy type appear similar to those on seedlings. Both waxy and wet shoots can be rooted, although the latter suffer some rotting during rooting. If shoot multiplication has been on SH medium, the last subculture is carried out on Gresshoff and Doy"s salts (1972) for preconditioning. Such methods allow for the production of 180-250 shoots per seed in ca. 12 weeks, when excised cotyledons are used (Aitken et al., 1981), but 20-30 weeks, depending on clone, are needed to obtain rootable shoots.

For rooting, single shoots are placed in water agar medium containing the auxins, indole butyric acid (4.92 μM) and naphthalene acetic acid (2.68 μM) for 5 days at 20°C. The treated shoots are then washed and planted in non-sterile peat-pumice mix in propagation trays under the same photoperiod and light intensity as that for shoot formation but at a lower temperature

20-25/15-20 (day/night) and at a high relative humidity (90-95%). Eighty percent of the shoots develop roots in 2-8 weeks. The roots of the plantlets thus formed are pruned back to 1-2 cm from the base of the shoot and replanted and fertilized. They are gradually hardened off by lowering the humidity, and grown-on in the glass house. The plantlets are then rootpruned a second time, and lined out in the nursery bed in late spring. They are ready for field planting the following winter.

At FRI micropropagation studies are also being undertaken with explants taken from mature trees. Two major problems have been encountered, namely obtaining tissue which is free of contaminants and which undergoes active growth in culture (Anon, 1983). To overcome these problems enclosed buds from branch tips are used. The hard scales of the buds are peeled off and they are cut into 2-10 pieces before being placed on a nutrient medium containing cytokinin. The number of shoots obtained per bud varies considerably, but up to 70 shoots have been obtained in 8 months.

Remultiplication capacity and cold storage of shoots are also being examined at FRI. Remultiplication of shoots is carried out by a topping method, in which the shoot apex and apical tuft of the primary needles (tops), are severed from each shoot. New shoots arise from the axillary meristems in the basal portions. These can be rooted in 3 months, remultiplied again or cold stored. The tops are grown further and either topped again or rooted. At least four fold remultiplication every 3 months is possible (Aitken-Christie and Thorpe, 1984; Smith, 1984a).

Rootable shoots can be placed in cold storage at 4°C under low light intensity (8_μ $Em^{-2}s^{-1}$) and a 16 hr photoperiod. The requirement for light, frequency of medium renewal, etc, have not been fully investigated as yet (Aitken-Christie and Gleed, 1984). To date shoots maintained in cold storage for 4.5 years appear healthy, but plantlet formation so far has only been carried out with material kept for 17 months in storage. The aim of this effort is to be able to form healthy plantlets from shoots kept in cold storage for 6-8 years, during which time material from the same clones would be under field evaluation. The best clones based on field performance could then be rapidly multiplied and the poor clones discarded. This approach may be valuable in reducing the problems associated with slow and low rates of propagating older material. It is also a possible alternative to hedging of pine grafts or cuttings and for germplasm storage (Aitken-Christie and Gleed, 1984). In addition, cold storage methods utilize less space and are disease free compared to hedges. However the fidelity as well as the rooting capacity of long term cold-stored shoots await the results of research now in progress.

At FRI cold storage is also used in a routine way, when space is limiting in the culture room, when labour is in short supply for transfers, or during holiday periods, etc, facilitating easier laboratory management (Aitken-Christie and Gleed, 1984).

LARGE SCALE PRODUCTION OF MICROPROPAGATED PLANTLETS

In 1979 a small production laboratory was established at FRI. The primary function was to use the methods of Horgan and Aitken (1981) and subsequent modifications as described above to produce planting stock for field trials, and to collect data on the cost of production. The actual methods used are described in detail elsewhere (Smith, 1984a).

In these studies utilizing embryos and excised cotyledons an average of 257 shoots per clone were obtained in 6 months, utilizing 65 clones (Aitken-Christie,unpublished; Smith et al.,1982; Smith,1984a). Provided the shoots were protected from infection, heat and water stress, rooting percentages of 80% or better within 8 weeks were achieved. Some clones gave as high as 95% rooting. Survival of rooted plants was found to be between 85 and 95%. Early field trials indicated that the growth rate of plantlets was the same as seedlings from the same seed source. Plantlets formed secondary needles earlier in the season than seedlings and had a reduced tendency to exhibit the open growth bud form characteristic of seedlings (Smith,1984a).

The cost of each step in plantlet production under the above conditions has been determined (Smith 1984a). This allows for an overall cost-benefit analysis, an important determinant for the use of micropropagation technology. Labour was found to account for 80% of the total cost of producing the planting stock. The figure is based on the assumption that known clones would be remultiplied in sterile culture, as net labour productivity of up to 90 shoots per hour at the sterile transfer bench could be obtained. If, however new seed from controlled crosses was introduced into the micropropagation programme labour productivity could drop to about 35 shoots per hour of labour. It is therefore apparent that a conflict could arise between the goals of the micropropagation laboratory manager and the tree breeder being serviced (Smith,1984b).

With present techniques and taking into account normal losses, etc, a price range of NZ$ 300 to $1000 per thousand plants was calculated (Smith, 1984a). For large scale operation $450 per thousand was considered a realistic figure. The cost for seed orchard seedlings is ca $45 per thousand. Nursery cuttings range from about $70 to $140 per thousand depending on the source and method used (F.R.I. unpublished). In contrast cuttings for seed orchard can be as costly as $7000 per thousand. Acceptable costs for planting stock depend on the expected value of the end product, the growth rate increment obtained and the initial planting density. On the basis of the current price range for standing timber, presently available micropropagation techniques with juvenile explants of radiata pine would be economically attractive if low planting densities were used, and if volume growth increments of 10-20% more than normal seedling plant stock could be assured (Smith,1984a).

ROLE OF MICROPROPAGATION

By the end of 1984 over 8000 tissue culture derived plantlets of radiata pine will be in the field at various sites at FRI in Rotorua, at Kaingaroa and at other sites in the major forested areas of the central plateau in New Zealand. This material is in various trials along with seedlings and cuttings selected from plants of various ages. It will be several years before evaluation data from the various replicated trials will be available. Some of the plantlets are also being used in physiological studies.

Two companies in New Zealand, Tasman Forestry Ltd. and Topline Nurseries Ltd, have recently become involved in the micropropagation of juvenile radiata pine. Topline, a 10 year old company, is involved in the vegetative propagation of ornamentals, fruiting and flower crops for domestic use as well as export. Presumably the company hopes to occupy a commerical niche for supplying plantlets of radiata pine for afforestation and agroforestry. Tasman Forestry Ltd is planning to utilize micropropagation as part of their effort towards clonal forest establishment on their lands. Tasman began an afforestation programme in 1962, and in 1972 the first seed orchard progeny was planted. Since then more than 30,000 ha of improved stock has been planted. As the resulting trees have better form and grow faster, they are cheaper and more profitable to manage, and are expected to produce operational cost advantages of NZ $371/ha (Aitken-Christie and Gleed 1984). These benefits have prompted the decision to establish a micropropagation unit.

Micropropagation of juvenile radiata pine is presently envisaged as having roles in six major areas of activity in research and industry (Aitken-Christie and Gleed,1984). These areas are: 1) cold storage of shoots in vitro, 2) micropropagation research, 3) research plantations, 4) production plantations, 5) introduction of new material and 6) in vitro selection.

The long term cold storage of juvenile clones in culture could facilitate clonal forestry on a large scale. As indicated earlier this will allow for field evaluation of clonal material and a rapid subsequent multiplication of tested clones, hopefully without ageing of the material used. A second use will permit clonal material to be used comparatively for studies on juvenile micropropagation itself, for leads in micropropagation of mature radiata pine tissues and as a research tool in physiology. The micropropagation of mature radiata pine is New Zealand is expected to play an important role in the production of trees for seed orchards, is an important tool for use in tree improvement research, and possibly could be used for clonal afforestation if rejuvenation is achieved (Aitken-Christie and Gleed,1984).

Research plantations with micropropaged stock may play an important role in evaluating field performance. To date no information is available on the long-term functioning of any tissue culture plantlets in the field. A logical extension to the above is the development of production plantations with improved propagules. A major problem in capitalizing on superior full-sib families has been the development of methods for carrying out controlled pollination on a scale large enough to produce a sufficient quantity of seed. Novel solutions to this problem have been proposed (Sweet and Krugman,1977), but these in turn have run into snags (Smith et al.,1981). However, more recently, controlled pollination using the bag isolation technique has begun for commerical production of full sib seed. Such seed will become available in 2 years, but it will be at least 8 years before commercial quantities of control pollinated radiata pine seed are available in New Zealand to provide for 10% of the total annual seed requirement (M. Carson, pers. comm.). Thus vegetative propagation has the opportunity to contribute through the mass clonal multiplication from superior seed. To this end Tasman Forestry Ltd working in close association with the F.R.I. plan to produce 250,000 micropropagated plantlets in 1985 and a further 2 X 10^6 plantlets in 1987 from control pollinated seed for field planting (Aitken-Christie and Gleed 1984).

Because of its high multiplication rates, micropropagation has been useful in the rapid introduction of new ornamental varieties (Murashige, 1978). A similar role can be envisaged in speciality types in radiata pine. "Taradale" is a new registered variety which exhibits a hedged growth form, and has potential as a shelter belt tree (Aitken-Christie and Gleed,1984). Micropropagation of this type could produce the plantlets for field trials and possibly for commercial production. Other similar uses could be imagined, e.g., rapid multiplication of stock resistant to the fungus Dothistroma pini.

In vitro selection has been used to select herbicide and disease resistant and salt tolerant cell lines in agricultural and horticultural crops and may play a similar role in radiata pine in the future. Research is needed to show whether important forest tree traits e.g., growth rate can be selected for under in vitro conditions.

THE FUTURE

Improvement in existing micropropagation techniques is an area of major concern at F.R.I. A number of stages where the present process can be improved have been identified (Smith,1984a). For example, rooting and growing on in containers suitable for later direct field planting will remove some of the handling and nursery-tending operations. Preliminary investigations have also shown that the last sterile culture stage, i.e. the root initiation stage in the presence of auxin, can possibly be eliminated by adding auxin, directly to the cultures during the last week of the pre-conditioning or multiplication stage (Smith,1984a). At least two transfers are presently needed to produce useful shoots. Limited preliminary work has been also shown that radiata pine shoots will grow on a polyurethane support medium with liquid nutrient supplement (Smith, unpublished). This opens up the possibility of a sterile hydroponic culture system, but such a system is not yet on the horizon. Even if successful, inexpensive methods for providing individual shoot initials at the beginning of a culture cycle is required, as this step is costly in terms of labour (Smith,1984a). It is clear that currently available micropropagation methods for radiata pine can be refined further, but it is not evident that a dramatic decrease in costs can be realized in this labour intensive method, even though the present techniques are attractive to some sectors of the industry.

Interests in alternative approaches to plantlet formation is thus high and work on somatic embryogenesis in radiata pine is being undertaken at the F.R.I. and in Calgary. Somatic embryogenesis in which cells, tissue and explants give rise to structures similar in appearance to normal zygotic embryos offers potential for producing very large numbers of plantlets in a less labour intensive process (Thorpe,1982; Smith,1984b). While somatic embryogenesis has been achieved in many herbaceous species, very few woody plants and no conifers have yet given rise to plantlets via this method (Evans et al.,1981; Biondi and Thorpe,1982; Thorpe and Biondi, 1984). Methods for dealing with somatic embryos by encapsulation to form artificial seed or for direct drilling in a gel formulation are under study (Murashige 1978; Sharp et al.,1982).

Other methods of vegetative propagation, including stem and fascicle cuttings are cheaper than micropropagation via organogenesis, however, multiplication rates are usually lower (M. Menzies, pers. comm.). Success with somatic embryogenesis will likely be more than cost competitive with the most efficient cutting programme and also posses other advantages.

9

Genetically superior radiata pine seed is a valuable New Zealand forest product (Aitken-Christie & Gleed,1984). Seed orchard seed is already being sold overseas and there are likely to be markets in countries like western U.S.A, Australia, Spain, Chile, Italy, Greece and South Africa for the controlled pollinated seed. In all probability not enough seed will be available even for New Zealand's own needs, thus,some form of vegetative multiplication is envisaged to make the seed go further. Micropropagation could be used to rapidly bulk-up superior families, thus creating an additional forest product, which will be easy to export because of its aseptic state.

Plant tissue culture techniques in general are of tremendous potential value to forest tree improvement (Thorpe,1983). The technology is envisaged as playing a complementary role to more traditional methods through exploiting spontaneous or induced genetic and epigenetic variability in culture, by use of haploidy and by use of protoplasts. Both haploids and protoplasts can aid in shortening breeding cycles and allow for unconventional crosses respectively. However under New Zealand's conditions, because of the long established and successful radiata pine breeding programme, the need for these alternative tree improvement approaches is not as readily apparent as in forest tree improvement programmes elsewhere. Nevertheless work with protoplasts has begun in relation to studies on somatic embryogenesis (Smith,1984b).

The combination of climate, radiata pine, research and management experience has made forestry in New Zealand very successful. The tree improvement programmes have produced superior trees, the management techniques have shown that plantation forestry can be profitable. Agroforestry on productive sites potentially offers even higher returns (Sutton,1984). The probability of a wood surplus in New Zealand is therefore real. The situation requires that the country identify where the markets will be or what wood products or qualities offer the highest returns and best prospects. Furthermore,New Zealand must do something soon to ensure the profitable and orderly marketing of the increased volumes of wood that will be available for export (Sutton,1984). The precise role of micropropagation in forest tree improvement and forest management in New Zealand under these conditions remains to be clearly defined. Nevertheless it is evident that a continued role exists in both research and industry.

ACKNOWLEDGEMENT

This article was written while T.A.T. was a Visiting Scientist in the Production Forestry Group of the Forest Research Institute, Rotorua, New Zealand in August 1984. The receipt of a Natural Sciences and Engineering Research Council of Canada Collaborative Research Grant, which made the visit possible, is gratefully acknowledged.

REFERENCES

Anon. 1979. N.Z. Forest Research Institute Annual Report, pp. 18.
Anon. 1983. Propagation and early growth. N.Z. Forest Research Institute Annual Report, pp. 21-25.
Aitken, J., K.J. Horgan and T.A. Thorpe. 1981. Influence of explant selection on the shoot-forming capacity of juvenile tissue of Pinus radiata. Can. J. For. Res. 11: 112-117.

10

Aitken-Christie, J., K. Horgan and D.R. Smith. 1982. Micropropagation of radiata pine. Mikroformering of skovtraeer. Ugeskr. Jordbrug 19: 375-382.

Aitken-Christie, J. and J.A. Gleed. 1984. Uses of micropropagation of juvenile radiata pine in New Zealand. In: Recent Advances in Forest Biotechnol. Mich. Biotech. Inst., East Lansing, in press.

Aitken-Christie, J. and T.A. Thorpe. 1984. Clonal propagation: Gymnosperms. In: Cell Culture and Somatic Cell Genetics of Plants, Vol. 1. Laboratory Procedures and their Application (I.K. Vasil, ed.), pp. 82-95, Academic Press, New York.

Biondi, S. and T.A. Thorpe. 1982. Clonal propagation of forest tree species. In: Tissue Culture of Economically Important Plants (A.N. Rao, ed.), pp. 197-204, COSTED & Asian Network of Biological Sciences, Singapore.

Burdon, R.D. 1982. The roles and optimal place of vegetative propagation in tree breeding strategies. In: Proc. IUFRO Joint Meeting of Working Parties on Genetics about Breeding Strategies including Multiclonal Varieties, Sensenstein, Sept. 6-10, pp. 66-83.

Evans, D.A., W.R. Sharp and C.E. Flick. 1981. Growth and behaviour of cell cultures: Embryogenesis and organogenesis. In: Plant Tissue Culture: Methods and Applications in Agriculture (T.A. Thorpe, ed.), pp. 45-113, Academic Press, New York.

Gleed, J.A. 1983. Tree improvement-first results from a radiata pine production forest. Appita 36: 386-390.

Gresshoff, P.M. and C.H. Doy. 1972. Development and differentiation of haploid Lycopersicon esculentum (tomato). Planta 107: 161-170.

Horgan, K.J. 1982. The tissue culture of forest trees. In: Proc. 9th Int. Symp., Natl. Acad. Sci., Republic of Korea, pp. 105-120.

Horgan, K.J. and J. Aitken. 1981. Reliable plantlet formation from embryos and seedling shoot tips of radiata pine. Physiol. Plant. 53: 170-175.

Murashige, T. 1978. The impact of tissue culture on agriculture. In: Frontiers of Plant Tissue Culture 1978, (T.A. Thorpe, ed.), pp. 15-26, 518-524, Univ. of Calgary Press, Calgary, Canada.

Quoirin, M. and P. Le Poivre. 1977. Etudes de milieux adaptes aux cultures in vitro de Prunus. Acta Hortic. 78: 437-442.

Reilly, K.J. and J. Washer. 1977. Vegetative propagation of radiata pine by tissue culture: Plantlet formation from embryonic tissue. N.Z.J. For. Sci. 7: 199-206.

Schenk, R.W. and A.C. Hildebrandt. 1972. Medium and techniques for induction and growth of monocotyledonous and dicotyledonous plant cell cultures. Can. J. Bot. 50: 199-204.

Sharp, W.R., D.A. Evans and M.R. Sondahl. 1982. Application of somatic embryogenesis to crop improvement. In: Plant Tissue Culture 1982 (A. Fujiwara, ed.), pp. 759-762, Maruzen, Tokyo.

Shelbourne, C.J.A. and I.J. Thulin. 1974. Early results from a clonal selection and testing programme with radiata pine. N.Z.J. For. Sci. 4: 387-398.

Smith, D.R. 1984a. Micropropagation of forest trees: Pinus radiata in New Zealand as a model system. In: Micropropagation of Fruit and Forest Trees (Y.P.S. Bajaj, ed.), Springer-Verlag, Berlin and New York, in press.

Smith, D.R. 1984b. Tissue culture for clonal forestry: Present and future technologies. In: Forest Industries and Biotechnology, Proc. 16th Biotechnology Conference, Massey Univ., New Zealand, May 1984, in press.

Smith, D.R., J. Aitken and G.B. Sweet. 1981. Vegetative amplification: An aid to optimizing the attainment of genetic gains from Pinus radiata. In: Proc. Symp. on Flowering Physiology. (S.L. Krugman and M. Katsuta, eds.), pp. 117-123.

Smith, D.R., K.J. Horgan and J. Aitken-Christie. 1982. Micropropagation of Pinus radiata for afforestation. In: Plant Tissue Culture 1982 (A. Fujiwara, ed.), pp. 723-724, Maruzen, Tokyo.

Sutton, W.R.J. 1984. New Zealand experience with radiata pine. The H.R. MacMillan Lectureship in Forestry, Feb. 23, 1984, U.B.C., Vancouver, B.C. p. 19.

Sweet, G.B. and S.L. Krugman. 1977. Flowering and seed production problems-and a new concept in seed orchards. Invited special paper, Third World Consultation on Forest Tree Breeding, Canberra, Australia.

Sweet, G.B. and L.G. Wells. 1974. Comparison of the growth of vegetative propagules and seedlings of Pinus radiata. N.Z.J. For. Sci. 4: 399-409.

Thorpe, T.A. 1982. Callus organization and de novo formation of shoots, roots, and embryos in vitro. In: Application of Plant Cell and Tissue Culture to Agriculture and Industry (D.T. Tomes, B.E. Ellis, P.M. Harney, K.J. Kasha and R.L. Peterson, eds.), pp. 115-138, Univ. of Guelph, Guelph, Ontario.

Thorpe, T.A. 1983. Biotechnological applications of tissue culture to forest tree improvement. Biotech. Advs. 1: 263-278.

Thorpe, T.A. and S. Biondi. 1984. Conifers. In: Handbook of Plant Cell Culture, Vol. 2, (D.A. Evans, W.R. Sharp, P.V. Ammirato and Y. Yamada, eds.), pp. 435-470, Macmillan, New York.

Wilcox, M.D. 1983. Forestry. In: Plant Breeding in New Zealand (G.S. Wratt and H.C. Smith, eds.), pp. 181-193, Butterworths N.Z. Ltd.

Shoot primordia induction on bud explants from 12 - 15 year-old Douglas fir: the requirement for BA and NH_4NO_3 in dormant collections.

David I. Dunstan,
Gina H. Mohammed,
Plant Tissue Culture Research Group,
Kelowna Nurseries Ltd.,
P.O. Box 178, Kelowna, B.C., V1Y 7N5.
and
Trevor A. Thorpe,
Plant Physiology Research Group,
Department of Biology,
University of Calgary,
Calgary, Alberta, T2N 1N4.

Abstract
Improved parameters for the induction of shoot primordia have been achieved for bud explants from dormant collections of 12 - 15 year-old Douglas fir. The improvements resulted after the augmentation of the Boulay medium B (1) with BA (1 mgl^{-1}) and NH_4NO_3 (400 mgl^{-1}). These results are reported in the context of developing a tissue culture propagation technique for the eventual production of true-to-type elite Douglas fir.

Introduction

Douglas fir (Pseudotsuga menziesii) is a commercially important forest tree in Western Canada, being mainly used in construction because of its durability and strength. However, due to the intensity of logging, its numbers are being seriously affected. The lack of appreciable seed production before age 20 - 25 (2) has seriously impaired genetic improvement by traditional means (3, 4, 5). There is also no commercially feasible traditional vegetative propagation technique. Plant tissue culture propagation may therefore have a valuable role to play in the provision of cloned elite trees. The development of protocols for the clonal tissue culture propagation of a number of plant species (6, 7) indicates the potential that such strategies could have.

To produce true-to-type elite trees vegetative explants from mature donor trees must be used. This is because seed explants from elite trees do not have the same genotype as the parents, and because elite characteristics cannot be accurately judged in immature trees.

There is, however, no reliable procedure for the production of tissue-cultured plants from such vegetative sources, although preliminary results have been obtained for Pseudotsuga menziesii (1, 8).

We report here findings relative to the NH_4NO_3 (ammonium nitrate) and BA (benzyl adenine) requirements for Douglas fir bud explants collected in the dormant season. This study forms part of a comprehensive investigation to systematically evaluate each sequential culture stage, with the construction of a commercially workable protocol for the true-to-type multiplication from 12 - 15 year-old Douglas fir.

Materials and Methods

Plant Material

Dormant (collected from 5th December 1983 to March 15 1984) vegetative buds were collected from 12 to 15 year-old Douglas fir in a natural forest stand in the Penticton Forest District of southern interior British Columbia. To do this portions of branches were harvested from the lower two-thirds of the tree, from the outer one-third to one-half of each branch.

Buds were dissected to remove their apical portions with the lower portion being used as the final explant.

Media

The nutrient formulation, B, described by Boulay (1) was used as the standard reference medium, with the difference that $AlCl_3$ was omitted. Agar (Sigma Chemical, St. Louis, Mo.) was supplied at 5 gl^{-1}.

The medium was compared to two modifications, supplemented with

$$\text{(a)} \quad \text{BA at 1 mgl}^{-1} \qquad = \text{B1}$$
$$\text{or} \quad \text{(b)} \quad \text{BA at 1 mgl}^{-1}$$
$$NH_4NO_3 \text{ at 400 mgl}^{-1} = \text{B2}$$

The three media were dispensed into 150 x 25 mm test tubes, 20 ml per tube, capped with Kaput closures (Bellco Glass, Vineland, N.J.), and autoclaved. Five buds were inoculated on each medium per repeat, and a minimum of two repeat passages were prepared.

Growth Conditions

Cultures were incubated at $25^\circ \pm 2^\circ$ C under an illuminance of 5.5 klx during a 16 hour light period. Cultures were transferred to fresh nutrient medium each 3 weeks. Explants that were inoculated onto media that contained BA were transferred to the same medium, without BA, at 6 weeks (the end of the second passage).

14

Data Collection

The following parameters were used to assess culture response:

(1) survival, measured from 6 weeks.
(2) percentage of original explants that showed the induction of shoot primordia.
(3) average number of induced primordia per explant (parameters 2 and 3, measured at 20 weeks).
(4) fresh and dry weights (10 explants from each medium harvested each 3 weeks, from 0 to 15 weeks).

For fresh weight determination, explants were blotted and weighed on foil squares. Subsequently the samples were air-dried overnight, prior to 24 hours drying in an oven, at 100° C. Dried samples were stored in a dessicator until weighing.

Results

Survival (monitored from 6 weeks) was similar for each medium (70 - 82%) (Table 1). Induced explants developed visible nodular primordia from the 6th week. These developed into recognizable shoot primordia (with apical dome and needle structures) from the 9th week. Primordia were produced, and developed, asynchronously over a long period of time. The effect of BA can be seen by comparison of the Boulay medium, B, with the BA-containing medium B-1. Explant induction on B was only 2% of surviving material, while it was 52% on B-1.

Table 1

Growth parameters comparing the response of dormant bud explants from 12 - 15 year-old Douglas fir, grown on three nutrient media.

				Medium*	
	Parameter	Age	B	B-1	B-2
A.	Survival	6 wk.	82%	76%	70%
B.	Explant Induction (% of A)	20 wk.	2%	52 %	86%
C.	Average number of primordia/expl.	20 wk.	2	7	9

*B   = Boulay, medium B (1)
 B-1 = B, supplemented with BA @ 1 mgl^{-1}
 B-2 = B-1, supplemented with NH_4NO_3 @ 400 mgl^{-1}

Further, the average number of primordia per induced explant was only 2 on the BA-free medium B, but was 7 on B-1. There was, therefore, an improvement both in the quantity of the initial bud explants that were induced and in the quantity of primordia that were formed when BA was present in the medium for the first 6 weeks of culture. Periods of time less than this were not as effective.

Initial results from comparisons of the Boulay medium B with those of Lepoivre (9) and the WPM (10) indicated that the latter NH_4NO_3-containing (400 mgl^{-1}) media stimulated better induction of shoot primordia. Consequently, a comparison of these media with that of Boulay supplemented with 400 mgl^{-1} NH_4NO_3 was made. The modified Boulay medium outperformed the Lepoivre and WPM in the induction of shoot primordia.

The combination of 400 mgl^{-1} NH_4NO_3 with BA (1 mgl^{-1}), medium B-2, resulted in a greater improvement of both explant induction (86%) and in the average number of shoot primordia per explant (9) (Table 1), when compared to medium B or B-1.

By the 6th week on the BA-free medium, B, there was visible explant axis and/or needle elongation. Where primordia were induced they were in apparent axillary positions (Figure 1). Once primordia were induced, axis elongation slowed or ceased. At this time (6 - 9 weeks) axes were approximately 1 cm. The recorded primordia induction for medium B occurred for a collection made on March 15th.

Figure 1 8X. Bud explant grown on medium B for first 6 weeks. Lateral production from apparent axillary positions, with prior axis elongation.

On the BA-containing media B-1 and B-2, explant size had increased to about 4 to 5 mm by week 9, but stabilized at the time of primordia induction (from week 9) and during shoot development from primordia. Two major patterns of induction were observed; primordia were induced either in apparent axillary positions (Figure 2), or by modified needle development (one or more primordia per needle) (Figure 3). Both patterns occurred to an equal degree.

No significant differences could be determined between fresh and dry weight data collected for each medium. There was generally a steady gain in both parameters, with dry weight being approximately 10% of the fresh weight.

The development of primordia into recognizable shoots could occur on all three media. However, the development was superior in cultures that were initiated on B-1 during the March collection.

Remultiplication from surviving excised shoots occurred following the removal of their apical portions, with the stimulation of one or two lateral shoots distally.

No rooting of excised shoots has yet been attempted.

Discussion

The addition of the cytokinin, BA, to the Boulay medium stimulated an improvement in explant induction and in the number of induced primordia per explant using dormant bud collections. Although cytokinins (CK) are present in low concentrations during dormant collections, they increase during the onset of the spring-flush (11). The requirement for exogenous BA in dormant collections of Douglas fir may complement the low endogenous level in achieving the correct threshold to elicit the induction response. Correspondingly, the induction of buds on medium B in the final collection of the study (March 15th) may reflect a gradual increase in endogenous cytokinin prior to renewed spring activity (bud-burst was recorded early in June).

The NH_4NO_3-supplemented medium, B-2, showed further improvements in the induction parameters. In a similar pattern to the observed increases in endogenous CK in the spring, Proebsting and Chaplin (12) revealed that the total nitrogen content of Douglas fir seedling trees was at a steady, but low, level in the fall and winter but increased in the spring. The requirement for exogenous NH_4NO_3 for improved explant induction parameters may therefore be analogous to their requirement of BA. In both situations it may be anticipated that the exogenous supplements would become less necessary as the respective endogenous levels rise in the spring. Further the requirement for these and other supplements could alter throughout the year. In these experiments no attempt has been made to analyze the requirements for the continued development of primordia into excisable shoots. It was observed, however, that high induction levels on B-2 did not always result in high level of development. This indicates that a different maintenance medium may be necessary. This appears to be in general agreement with Boulay (1).

Figure 2   5X.   Bud explant grown on medium B-1 for 6 weeks.
            Lateral production from apparent axillary positions,
            without prior axis elongation.

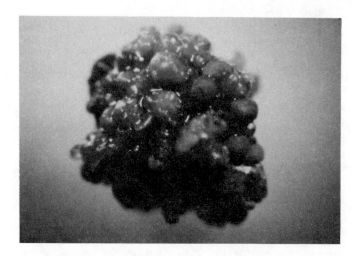

Figure 3   12.5X.   Bud explant grown on medium B-2 for 6 weeks.
            Primordia production by modified needle development
            without prior exis elongation.

18

The two observed patterns of primordia induction on BA-containing media do not appear to be related to the presence or absence of exogenous NH_4NO_3. At the present time it is not possible to determine what promotes each type of response, or how the two patterns develop. This is the subject of an ongoing microscopy investigation.

The overall aim of these experiments is to analyze the cultural requirements for the optimal induction and continued growth of primordia. At the present time it has been possible to obtain superior induction parameters for dormant bud collections by augmentation of the Boulay medium B (1) with BA and NH_4NO_3. Ongoing studies will determine the media requirements for other collection periods, and those that are necessary for continued optimal development of primordia into excisable shoots.

References

1. Boulay, M. 1979. Propagation in vitro du Douglas par micro-propagation de germination aseptique et culture de bourgeons dormants. In Micropropagation d'Arbres Forestiers, Annales AFOCEL, No. 12, 6/79. pp. 67 - 75.

2. Puritch, G.S. 1972. Cone Production in Conifers. Canadian Forest Service Information Report BC-X-65. 94 p. In R.P. Pharis, S.D. Ross and E. McMullan. 1980. Promotion of Flowering in the Pinaceae by Gibberellins. III. Seedlings of Douglas Fir Physiol. Plant. 50: 119 - 126.

3. Allen G. and Owens, J.N. 1972. The Life History of Douglas Fir. Information Canada, Ottawa. 139 p.

4. Pharis, R.P., Ross, S.D. and McMullan, E. 1980. Promotion of Flowering in the Pinaceae by Gibberellins. III. Seedlings of Douglas Fir. Physiol. Plant. 50: 119 - 126.

5. Thorpe, T.A. and Biondi, S. 1984. Conifers. In W.R. Sharp, D.A. Evans, P.V. Ammirato and Y. Yamada, eds., Handbook of Plant Cell Culture, Vol. 2: Crop Species. MacMillan (New York). pp. 435 - 470.

6. Hughes, K.W. 1981. Ornamental Species. In B.V. Conger, ed., Cloning Agricultural Plants Via In Vitro Techniques. CRC (Boca Raton). pp. 5 - 50.

7. Dunstan, D.I. and Thorpe, T.A. 1984. Plant Tissue Culture Technology and its Potential for use with Forest and Bioenergy Tree Species. Fifth Bioenergy R & D Seminar, Ottawa.

8. Thompson, D.G. and Zaerr, J.B. 1981. Induction of Adventitious Buds on Cultured Shoot Tips of Douglas Fir. In IUFRO Int'l Workshop on In Vitro Cultivation of Forest Tree Species. AFOCEL (Fontainebleau). pp. 167 - 174.

9. Quoirin, M., Ph. Lepoivre and Ph. Boxus. 1977. Un premier bilan de 10 annees de recherches sur les cultures de meristemes et la multiplication _in vitro_ de fruitiers ligneux. C.R. Rech. 1976 - 1977 et Rapports de Synthese, Stat. Cult. Fruit. et Maraich. Gembloux. pp. 93 - 117.

10. Lloyd, G. and McCown, B. 1980. Commercially Feasible Micro-propagation of Mountain Laurel, _Kalmia Latifolia_ by use of Shoot Tip Culture. _In_ I.P.P.S. Combined Proceedings, Vol. 30. pp. 421 - 427.

11. Ross, S.D., Pharis, R.P. and Binder, W.D. 1983. Growth Regulators and Conifers: Their Physiology and Potential Uses in Forestry. _In_ L.G. Nickell, ed., Plant Growth Regulating Chemicals, Vol. II. CRC (Boca Raton). pp. 35 - 78.

12. Proebsting, W.M. and Chaplin, M.H. 1983. Elemental Content of Douglas Fir Shoot Tips: Sampling and Variability. Commun. Soil Sci. and Plant Anal. 14 (5): 353 - 362.

IN VITRO PROPAGATION OF MATURE CONIFERS

J.M. Bonga, Canadian Forestry Service, Maritimes Forest Research Centre, P.O. Box 4000, Fredericton, N.B., E3B 5P7, Canada

INTRODUCTION

Considerable improvement of conifer planting stock has been achieved by breeding (Farnum et al. 1983), and further improvement can be expected because of the large, untested and unused gene pool that is still available. However, breeding is a long and difficult process because 1) it takes many years from seed to sexual maturity of the tree, 2) conifers are mostly extremely heterozygous and, therefore, are subject to inbreeding depression, and 3) many of the desired traits are determined by nonadditive gene combinations, breeding will generally lead to a breakup and loss of these combinations. Therefore, many foresters are considering clonal propagation of selected, superior specimens (plus trees) as an alternative to conventional breeding.

The main advantages of clonal propagation of superior trees are 1) in contrast to stock improvement by breeding, gains obtained by clonal propagation are immediate, and 2) favorable nonadditive gene combinations are transmitted intact to the propagules. Therefore, compared with the original tree population, or first generation half sib or full sib breeding populations, the genetic gain in clonal populations can be considerable. In Eucalyptus, for example, which is clonally propagated on a large scale, volume gains over those of the original populations have been reported as high as 54% by sexual breeding and 100% for clonal populations (Hartney 1983, Zobel and Talbert 1984). In conifers, the potential for gain by cloning may not be quite as large as in Eucalyptus, but should still be considerable. However, this does not apply to all conifers. There are a few species e.g., Pinus resinosa that have a relatively narrow genetic base, and therefore, do not have distinct plus trees.

One of the major problems in clonal propagation of conifers is that trees old enough to have expressed their full potential, i.e., trees selected at about half the rotation age, are generally difficult to clone. Rooting of cuttings, the most common traditional method of clonal propagation, and in vitro propagation work well only with juvenile material. Of these two techniques, in vitro propagation probably offers the best prospects of eventual success because 1) in vitro propagation is carried out under aseptic conditions and consequently, chemicals that would degrade rapidly under the nonaseptic conditions used for rooting of cuttings can be applied over long periods of time, and 2) in in vitro propagation small masses of tissue are used, and thus some of the internal, correlative controls that operate in such large units as stem cuttings are removed.

Over the years we have tried in vitro propagation of several conifer species (Abies balsamea, Pinus resinosa, P. banksiana, P. mugo, Picea abies, P. glauca, and Pseudotsuga menziesii)(Bonga, 1981), with little success. Highly stunted adventitious shoots, or abnormal embryolike structures were sometimes formed, but no normal propagules. Over the last few years we have switched our attention to Larix decidua. This species performs better in vitro, and large numbers of adventitious shoots are now routinely being produced. These experiments are described below.

MATERIAL AND METHODS

There are two types of explants that form adventitious shoots in vitro. These are female cones collected at the time of meiosis (Bonga 1984) and primordial shoots collected in August and September before the onset of dormancy.

The female cones and primordial shoots were obtained from Larix decidua trees 25-30 years old. The female cones were collected once or twice a week in late April and early May, i.e., around the time of meiosis, which generally occurs in the first week of May. The cones were surface sterilized and then cut in 1- to 2-mm-thick transverse slices, which were placed on a Litvay et al. (1981) medium, with all minerals, except FeEDTA, reduced to half strength. In most cases, the initial medium contained 5×10^{-5}M benzylaminopurine (BAP) and no auxin. After 4-6 weeks the tissues were transferred every 2 weeks to fresh medium without BAP and auxin, and with or without 1% charcoal. The cultures were exposed to GROLUX fluorescent illumination at 50 $\mu Em^{-2}s^{-1}$ for 16 h daily, and were kept at constant 21°C.

Primordial short-, long-, female-, and male-shoots were excised from surface sterilized developing buds collected at weekly intervals throughout August and September. They were cultured on the same initial and subculture medium, and were exposed to the same illumination and temperature as the female cones collected at about the time of meiosis.

RESULTS

Female Cone Cultures

The experiments with female cones were carried out once a year over a four year period. In each of those years, the formation of adventitious shoots was restricted primarily to slices of cones collected at the time of meiosis. On slices of cones collected within one week before or after meiosis, they occurred occasionally; on slices of cones collected outside this period they did not occur. Some of the shoots appeared on the bracts or ovuliferous scales, but most of them originated from callus. Some of the callus produced adventitious shoots up to the 10th subculture of the callus.

The shoots developed mature-length needles and distinct apical, axillary, and lateral buds. The stems of the shoots either did not elongate or only grew a few millimetres in length. These shoots did not root.

Primordial Shoot Cultures

The experiments with primordial shoots were carried out during two consecutive summers. During both summers, shoots collected from late August till the end of September produced adventitious shoots, often in dense masses. These shoots were divided and subcultured. In subculture, they produced more shoots, mostly from callus, but also by lateral or axillary bud formation. In a few cases, the callus maintained its shoot-forming capacity up till the 10th subculture. Like the adventitious shoots formed from slices of female cones, the ones originating from primordial shoots formed distinct apical buds and needles of a length similar to that of mature needles in the field. The primordial long shoots had the highest capability for forming adventitious shoots, followed by the primordial female shoots. The primordial short shoots were less productive and the male shoots produced only a few adventitious shoots. In most shoots, stem elongation did not occur, but in some shoots a stem of several millimetres long, and in one case a stem 12 mm long was formed. Two shoots developed a distinct root spontaneously, i.e., without an auxin rooting treatment. These roots had a bright red tip and grew down into the agar medium. The longest one reached a length of 25 mm. The rooted shoots did not survive transfer to soil.

DISCUSSION

The experiments have shown that explants from mature trees of at least one conifer species, Larix decidua, have the capacity to form a large number of adventitious shoots, a very small percentage of which will root. Of course this is far from an effective mass in vitro propagation system for these trees. Lack of stem elongation, low rooting rates, and failure to survive transfer to soil are major obstacles still to be overcome.

REFERENCES

Bonga, J.M. 1981. Organogenesis in vitro of tissues from mature conifers. In Vitro 17: 511-518.

Bonga, J.M. 1984. Adventitious shoot formation in cultures of immature female strobili of Larix decidua. Physiol. Plant., In press.

Farnum, P., Timmis, R., and Kulp, J.L. 1983. Biotechnology of forest yield. Science 219: 694-702.

Hartney, V.J. 1983 Tissue culture of Eucalyptus. Comb. Proceed. Intern. Plant Propagators' Soc. 32: 98-109.

Litvay, J.D., Johnson, M.A., Verma, D., Einspahr, D., Weyrauch, K., 1981. Conifer suspension culture medium development using analytical data from developing seeds. Inst. Paper Chem. Tech. Paper Ser. 115: 1-17.

Zobel, B. and Talbert, J. 1984. Applied Forest Tree Improvement. John Wiley & Sons, New York.

BIOTECHNOLOGY IN BREEDING FOR BIOMASS PRODUCTION

Rong H. Ho
Ontario Tree Improvement and Forest Biomass Institute
Ministry of Natural Resources
Maple, Ontario, Canada

and

Louis Zsuffa
Faculty of Forestry
University of Toronto
Toronto, Ontario, Canada

Abstract

Some constraints in conventional tree breeding hinder progress in developing high yielding varieties. New techniques in biotechnology and molecular biology may remove some of the crossing barriers and obstacles in breeding and cloning. This article discusses the potential for the use of in vitro pollination and fertilization, in ovulo embryo culture and embryo rescue, protoplast fusion, dihaploids, tissue culture-induced variants, and genetic manipulation in tree breeding for biomass production.

Introduction

Crossing of trees with different genetic characteristics often leads to progenies or hybrids with better growth and development than shown by the parental trees. The progenies from wide crossing and the production of hybrids are not only an important means to augment a gene pool but also an important tool to improve tree species.

Many crop species are interspecific or intergeneric hybrids and contain different genomes. The hybrids have been either arisen from spontaneous crosses or produced by the breeders. Some are obtained with the aid of in vitro pollination and hybridization, in ovulo embryo culture and embryo rescue, or protoplast fusion to overcome sexual incompatibility.

Most tree species have great genetic variability facilitating improvement through intraspecific crossing or interspecific hybridization. Some lack intraspecific diversity, and improvement of the species can best be achieved by interspecific or intergeneric hybridization. However, the hybridization is often hampered by prefertilization or postfertilization incompatibility, and results in the failure to produce viable seed. Application of the in vitro techniques may provide us with new hybrids which cannot be produced by sexual reproduction for tree improvement.

Many generations of inbreeding or selfing to produce homozygous lines followed by outcrossing and recurrent selection have been very successful in the improvement of crop species. However, the usefulness of these techniques in tree species is not known because no homozygous line is available to be used for assessment. Recent advances in haploid induction have overcome the impracticality in conventional methods. Isogenic pure lines can be produced after diploidization of haploids and used for breeding and recurrent selection after reaching maturity.

This article discusses the potential for the use of in vitro pollination and fertilization, in ovulo embryo culture and embryo rescue, protoplast fusion, dihaploids, tissue culture-induced variants, and genetic manipulation in tree breeding for biomass production.

In Vitro Pollination and Fertilization

In vitro pollination and fertilization is most useful when the pollen tube fails to deliver the gamete to fertilize the egg or where early abscission of flowers occurs before maturation of the seeds. Although the technique has the potential to bypass the prefertilization barriers, its usefulness depends on the ability to culture the fertilized egg and ovule to a mature seed with full embryo.

The prefertilization barriers which prevent fertilization include gametophytic and sporophytic incompatibility. Gametophytic incompatibility results from the impediment or arrest of the growth of the pollen tube, and the gamete is not delivered prior to degeneration of the egg. Sporophytic incompatibility is expressed as the failure of the pollen to germinate on the stigma or the pollen tube to penetrate through the stigma. The barriers can sometimes be overcome by modifying the floral structure or using the mentor pollen to facilitate the completion of fertilization. The failure of pollen germination or pollen tube penetration may be corrected by the application of chemicals, and premature floral abscission may be prevented by spraying the flower with plant growth regulators.

In many plant species, these barriers cannot be overcome by the in vivo methods, and in vitro techniques have to be used to complete sexual reproduction. However, before the application of in vitro techniques, it is essential to have a good understanding of the reproductive processes and identify the barriers to viable seed production.

Kanta et al. (1962) were the first to successfully use in vitro pollination and fertilization in poppy (Papaver) species. The poppy anthers were collected at the time of anthesis and sterilized by exposure to UV light. The ovaries were dipped in alcohol and then flamed for sterilization. The ovules were excised, placed on a culture medium, and then pollinated. Pollen germinated in 15 minutes, and in 2 hours, the ovules were covered with pollen tubes. Full development of the embryo and endosperm occurred in 22 days. To date, about 15 interspecific and intergeneric hybrids have been produced using this technique.

Explant sources ranging from the whole flower to individual ovules have been used in the culture for in vitro pollination and fertilization. Success depends on the source and the developmental stage of the maternal tissue at the time of culture and may vary from species to species (Zenkteler 1980). Genotype of the pollen and pistillate parent also plays a very important role in the success of obtaining a hybrid in vitro (Gengenbach, 1977). Other important factors are medium manipulation, temperature control, pollen density in pollination, and the position where pollen is applied (Stewart, 1981).

Incompatibility mechanisms have prevented successful production of hybrids in many interspecific and intergeneric crosses in tree species. Gametophytic incompatibility has been reported in birch (Betula) and alder (Alnus) resulting from the arrest of pollen tube growth in the style (Hagman 1975), in poplars (Populus) and oaks (Quercus) resulting from the impediment of pollen tube growth in the style (Guries and Stettler, 1976; Piatnitsky, 1947), and in hard pine (Pinus, subgenus Pinus) and spruce (Picea) resulting from the retardation of the pollen tube growth in the nucellus (Kriebel, 1972; Mikkola, 1969). Incompatibility in tree species often blocks the extension of genetic variability beyond the limit of gene pool of a given species. This limitation will be more serious if genetic variability is lacking in a species. The use of in vitro pollination and fertilization could facilitate the production of new hybrids to augment a gene pool and to combine desirable traits to capitalize heterosis for either timber, paper and pulp wood, or biomass production.

In Ovulo Embryo Culture and Embryo Rescue

In ovulo embryo culture and embryo rescue are designed to assist progenies affected by postfertilization barriers resulting in embryo or endosperm breakdown. The causes include ploidy difference, chromosome alteration, chromosome elimination, genic or cytoplasmic incompatibility, and seed dormancy.

Embryo degeneration could occur in the proembryo or embryo stage while endosperm breakdown could take place at any stage of embryogenesis (Reed and Collins, 1980). These could contribute to the formation of an empty seed, a seed without an embryo, or a naked embryo without surrounding endosperm.

Since embryo or endosperm could break down at any stage of embryogenesis, the embryo has to be rescued for culture prior to its abortion. Therefore, it is necessary to have an a priori knowledge of the in vivo reproduction systems for the species of interest. In addition, the selection of either in ovulo embryo culture or embryo rescue depends on whether the maternal tissue or endosperm is vital or detrimental to embryo survival, or the ease of culture to procure viable seed.

In ovulo embryo culture can be accomplished by culturing an intact flower, ovary, placenta attached to ovule, or individual ovule (Beasley, 1977). Withner (1942) was the first to successfully culture the excised orchid (Cattleya species) ovules while ovary culture was done in tomato (Lycopersicon) and gherkin (Cucumis) by Nitsch (1951). Maheshwari (1958) made a major advance in culturing the ovule with two-celled proembryo to maturity in poppy species, and Stewart and Hsu (1977) in culturing the ovule with zygote to maturity in cotton (Gossypium). To date, about 50 interspecific and intergeneric hybrids have been obtained via in ovulo embryo culture.

Since the first embryo rescue to produce an interspecific hybrid of flax (Linum) species (Laibach 1925), about 80 interspecific and intergeneric hybrids have been procured over the years (Collins and Grosser 1984). Embryos can be rescued at any stage of embryogenesis. However, the success depends on the species, the embryo age, medium composition, and culture environment. A proembryo is heterotrophic in nature and requires metabolites from the surrounding endosperm. An elaborate medium manipulation would be necessary to accomplish the rescue. At the embryo stage, the young embryo with cotyledon initials is relatively autotrophic in nature and more or less independent of the endosperm for subsequent growth. The medium requirement for the embryo growth to maturity would be less complex.

A relatively complicated medium and technique have been used in culture of conifer proembryos. Radforth and Pegoraro (1955) cultured pine proembryos and they grew to various sizes and groups of embryonic cells. Thomas (1972) isolated pine proembryos at the moment of cleavage and cultured them on a medium in the presence of nurse colonies of callus tissue. The proembryos had a comparable development to the embryo growth in situ in suspensor cell elongation and the embryonic cell division. Limited success has also been reported in ovule culture of the entire eastern white pine (P. strobus) megastrobili (Kriebel, 1975).

There is a general pattern that proembryos degenerate before entering the gametophyte or endosperm in postfertilization incompatibility in tree species. In the interspecific hybridization of soft pine (subgenus Strobus), degeneration of proembryos occurs at the 4-nucleate stage (Hagman and Mikkola, 1963) or after the embryo cleavage with 8 cells in the embryo initial (Kriebel, 1975). The proembryos of spruce hybrids degenerate at the 8-cell stage (Kossuth and Fechner, 1973) while degeneration in poplars occurs after zygote formation and the beginning stage of embryo development (Melchior and Seitz, 1968). Selfing to produce homozygous lines has been met with proembryo degeneration in both Douglas-fir (Pseudotsuga menziesii) (Orr-Ewing 1957) and white spruce (P. glauca) (Mergen et al. 1965). These indicate that the gametophyte or endosperm is either unable to nourish the developing embryos, or antagonistic to the developing embryos due to genic or cytoplasmic incompatibility.

The ability of the fertilized egg to develop from the zygote to the proembryo manifests the potential for growth to maturity. Isolation and culture of the proembryo before degeneration may produce a new hybrid which cannot be obtained through conventional tree breeding. The technique also has the potential to produce more than one viable embryo per ovule in conifers where crosses are made in compatible species. The isolation of proembryos for culture has to be carried out before degeneration of the proembryos in simple polyembryonic genera such as Picea and Pseudotsuga, and cleavage polyembryonic genera such as Pinus, Tsuga and Larix. The production of a new hybrid will augment the gene pool which is especially important for a species lacking in genetic variability and small in population size. The homozygous lines from embryo rescue following selfing may fulfill the tree breeders' dream for outcrosses and recurrent selection to capture heterosis.

Protoplast Fusion

The use of protoplast fusion has provided an unique opportunity to produce a hybrid which is unattainable through sexual reproduction. The technique offers the advantage of bypassing the prefertilization and postfertilization barriers. The hybrid would have both a combined genome and a novel cytoplasmic mixture.

Protoplast fusion is a physical, not a physiological process (Harms, 1983). Therefore, it is non-specific and the fusion can be carried out between any two species. However, the production of parasexual hybrids could only be produced between two compatible species in which the cell divisions are synchronized, morphogenesis is properly regulated, the genomes are balanced, and the interaction between genomes and cytoplasmic factors are well adjusted.

The production of hybrids from two fused somatic cells (parasexual or somatic hybridization) presents an amphidiploid (or allotetraploid) while regeneration of the plantlets from two fused reproductive cells (microspores or eggs) (parasexual hybridization) gives an amphihaploid (or allodiploid). The amphidiploid will be fertile and can be crossed with another amphidiploid of the same combination of fusion to produce a fertile progeny. However, if an amphidiploid ($N_1N_1N_2N_2$) is mated with another amphidiploid of a different species combination of fusion ($N_3N_3N_4N_4$), the progeny produced will have 4 different genomes ($N_1N_2N_3N_4$) and can only be made fertile after diploidization. If the amphidiploid is backcrossed to one of the original diploid parents, the product will be a triploid plant with 3 genomes where 2 of the genomes will come from the same species ($N_1N_1N_2$ or $N_1N_2N_2$). The amphihaploid is more likely to be sterile and has to be diploidized before it can be used in breeding.

Infertility in the hybrids will be a problem if vegetative propagation is difficult and seed production is the only means of propagation. For example, rooting of cuttings is not a problem in cottonwood and balsam poplar (Populus, section Aigeiros and Tacamahaca) while aspen (Populus, section Leuce) cuttings are hard to root. A somatic hybrid of cottonwood and aspen might provide us with an amphidiploid having the superior traits of both species. Backcrossing the amphidiploid to either cottonwood or aspen would produce a triploid with 2 genomes from either cottonwood or aspen. Triploid aspen has been reported to be superior to its diploid cousin (Johnsson, 1953). Ploidy level would not present a major problem in growth and productivity to broadleaf species since about one third of angiosperm species and many of the economically important species are polyploid. However, ploidy level will be a major concern in conifers since 4.6% of the gymnosperm species are naturally polyploid and only 3 polyploid species occur in conifers (Khoshoo, 1959). All artificial polyploids and most aneuploids in conifers have so far proven to be dwarf (Khoshoo, 1959). Darlington (1937) has theorized that the ratio between the chromosome number and the cell size has reached an equilibrium and any change in the ratio will be deleterious to growth and development of the species. If the hypothesis is correct, production of an amphidiploid between a soft and hard pine may not provide any promising hybrids. An amphihaploid from the two pines could be sterile and vegetative propagation would be the only means of propagation. However, percentage in rooting of cuttings in conifers is negatively correlated to the age of trees and a technique has to be developed to slow the maturation process. If the amphihaploid is diploidized, the resulting plant will be an amphidiploid. Therefore, parasexual hybridization might be of greater value in angiosperm tree species than in conifers.

Parasexual hybridization theoretically results in a formation of plants with both novel nuclear and cytoplasmic combination. However, some hybrids of distantly related species were found to be unstable, and chromosome elimination occurred at random and was observed in the subsequent transfer (Kao, 1977). The fate of the cytoplasmic organelles following fusion is also interesting. It appears that the hybrids favor mitochondrial recombination, but not recombination of the chloroplasts. The two chloroplast populations do not mix and apparently sort out to produce hybrids with one of the two parental chloroplasts (Boeshore et al., 1983).

Since the first report of successful protoplast fusion (Carlson et al., 1972), 26 interspecific and 6 intergeneric hybrids have been produced. Many other hybridizations have been attempted, and the fused cells have undergone division, but not regeneration of plantlets. Elm (Ulmus) species were used in one of the attempts. Although fusion was observed, there was no cell wall regeneration or cell division (Redenbough et al., 1981). The ability to regenerate plantlets from the protoplasts is a prerequisite for the success of somatic hybridization. Regeneration of a plantlet from protoplasts has not been accomplished in forest tree species. Therefore, this area of research requires immediate attention. The possibility of plant regeneration from protoplasts is within reach since colonies and calli have been produced (David and David, 1979) and plantlet regeneration has been achieved by in vitro vegetative propagation in tree species (Bonga 1977).

Dihaploids

Dihaploids are isogenic pure lines produced by diploidization of haploids. They are different from the homozygous lines produced after many generations of inbreeding or selfing since the latter still contain some degree of heterozygosity.

Naturally occurring haploid embryos have resulted from apomixis in nearly one hundred angiosperm species (Kasha, 1974). Many other species have been induced to produce haploid embryos by interspecific hybridization, pollination with irradiated pollen, and postpollination treatment with toluidine blue (Illies 1974). Androgenesis of the male gametophyte in pollination to produce haploid embryos have been reported in tobacco (Nicotiana), barley (Hordeum), and pepper (Capsicum) (Campos and Morgan, 1958). These methods of haploid embryo production were time-consuming and could not be relied on for regular and mass production. A reliable technique was not developed until Guha and Maheshwari (1964) reported the induction of haploids through anther culture. However, the technique did not arouse worldwide interest until some 5 years later as a result of a report by Nitsch and Nitsch (1969) on tobacco. To date, haploid plantlets have been induced through androgenesis in about 160 angiosperm species.

The induction of young pollen grains to proliferate and form haploid embryos or calli relies on a transformation in the genetic expression from gametophytic to sporophytic development in pollen grains by the culture conditions. The transformation in the pollen grains of poplar species by anther culture was reported by Wang et al. (1975) and the Heilungkiang Institute of Forestry, China (1975). By 1980, the culture had been successful in 14 poplar species (Zhu et al. 1980). In the meantime, horse chestnut (Aesculus hippocastanum) (Radojevic, 1978), birch (B. pendula) (Huhtinen 1978), rubber plant (Hevea brasiliensis) (Chen et al., 1979), P. maximowiczii x deltoides (Ho et al., 1983), and 6 other poplar hybrids and species (Ho, unpublished data) have been reported to be producing haploid plantlets through anther culture.

The procurement of isogenic pure lines is achieved by diploidization of the haploids with colchicine (Tanaka and Nakata, 1969) and by induction of endomitosis (Nishi and Mitsuoka, 1969). The pure lines have shown to be remarkably stable in meiosis thus enhancing their value in plant breeding (Collins and Sadasivaiah, 1972).

The use of dihaploids in variety selection and breeding is extensive in agriculture. A new variety of tobacco which is more resistant to bacterial wilt and has a mild smoking quality was developed by Nakamura et al. (1974). Three high yielding and superior varieties of tobacco have also been produced by scientists at the Chinese Academy of Agricultural Sciences (1978). Anther-derived dihaploids of rape (Brassica napus) are available and under field trials (Keller and Stringam, 1978). Considerable work in dihaploid breeding in rye (Secale cereale) and potato (Solanum tuberosum) has been initiated in West Germany (Wenzel, 1980).

The value of homozygous and isogenic lines is not known in forest trees. There are no homozygous lines for evaluation because of prefertilization and postfertilization incompatibility and the death of progenies due to inbreeding depression. The production of isogenic pure lines is in its infancy and their usefulness can be evaluated through progeny test only after the plants reach the reproductive age. The genetic expression of the two identical genomes in dihaploids eliminates the complexities of dominant alleles versus recessive alleles in diploids. The dihaploids will facilitate recessive gene recovery, and other genetic and physiological studies. Maximum heterosis in desirable trees may be achieved by outcrossing and recurrent selection as it has been done in many crop species.

Tissue Culture - Derived Variants

In the early stage of in vitro culture research, palingenetic expectation prevailed in using tissue culture as a means to clone a particular genotype. Each plant regenerated in vitro from a piece of tissue was expected to be an identical clone of the donor plant. However, many phenotypic variants were often observed in the produced plants and usually labelled as the artifacts of tissue culture. Recent evidence has shown that these variants may have either altered genomes or differential expression of the genomes.

Plant cell and tissue culture generates genetic variability. The variability is termed as somaclonal variation and the term somaclone is coined for the plants produced from any form of cell culture (Larkin and Scowcroft, 1981). The nature of the heritable changes in the plants regenerated from cultured cells or tissue is not known and has been attributed to the mutagenic culture environment. The changes include gross and cryptic chromosome alteration such as polyploidy, aneuploidy, chromosome rearrangement, transposition of transposons, and gene deletion or amplification.

Plant cell and tissue culture has provided an option to rapidly produce plants with increased genetic variability and has been praised as one of the most significant potential adjuncts to plant improvement. The potential usefulness of somaclonal variation was first realized in sugar cane (Saccharum) (Heinz & Mee 1969). The variations have been observed in morphology, cytogenetics, isozymes, and disease resistance. From somaclonal screening, a number of sugar cane varieties have been found to be resistant to either Fiji disease (virus) or Downy mildew (Sclerospora sacchari), and some somaclones were identified with increased resistance to both Fiji disease and Downy mildew (Krishnamurthi and Tlaskal 1974). Heins et al. (1977) found that somaclones varied greatly in their reaction to eyespot disease (Helminthosporium sacchari). Somaclonal variation has also been reported in potato, tobacco, rice (Oryza), maize (Zea mays), cauliflower (Brassica sp.), rape, and many other species.

Shepard et al. (1980) has also found a similar variation in plants regenerated from leaf protoplasts of potato and referred to as protoclones. The variation was observed over a period of 3 tuber generations and consistent changes were found in the tuber shape, yield, and maturity date, photoperiod requirements for flowering, plant morphology and disease resistance.

In forest trees, information on somaclonal variation has started to unfold. Lester and Berbee (1977) reported a wide range of variation in height, number of branches and leaf traits in poplar trees regenerated from a hybrid clone. Chromosome counts of root tips from many ramets of a somaclone were highly variable and they suggested that the reduced vigor of some ramets might be related to the instability of chromosome numbers. Cheng and Smeltzer (1981) found that loblolly pine (P. taeda) produced in vitro exhibited some variation in appearance. A large number of plants possessed long and thickened needles and the incidence of 4 needles per dwarf shoot (normally 3 needles per shoot) was much greater. Occasionally, a dwarf shoot was found to have 5 or 6 needles. Although the information on somaclonal variations is scanty at present, it is evident that the variations exist in forest tree species. These variations may be a problem for cloning and maintenance of a particular genotype by cell and tissue culture. However, they are a novel source for tree breeders who are looking for variants.

Genetic Manipulation

A plant protoplast is a naked cell which is potentially capable of cell wall regeneration, growth, and division. Protoplasts have been used as gametes for fusion to produce hybrids. A naked cell is also excellent material for genetic manipulation. Foreign genes could be introduced into the protoplasts by transduction, transformation, transcession, and invagination. The modified protoplasts are then regenerated to produce plants with the introduced genes. Doy et al. (1973) reported the successful transfer of the lac gene from bacteria into the cells of cress (Arabidopsis) and tomato by using transducing bacteriophages. Johnson et al. (1973) reported the similar success of introducing the lac gene into sycamore maple (Acer pseudoplatanus) cells. However, Malhotra et al. (1979) concluded that no lac gene was incorporated into the plant cell genome of Datura innoxia haploid callus by using bacteriophage lambda as a gene carrier. Extensive studies with pea (Pisum sativum), tomato, and barley also failed to confirm the presence of bacterial DNA in plant DNA (Kleinhofe et al., 1975).

Transplanting of the nucleus of a cell of one species for that of another species by microsurgery would result in a cell with a new combination of nucleus and cytoplasm. The plants produced from the restituted cells would facilitate the study of nucleus-cytoplasm interaction and the function of the nucleus and cytoplasm. For example, a plant with hard pine cytoplasm and soft pine nucleus, or vice versa, would present information on the interaction and might provide a bridge for interspecific hybridization between the two subgenera.

Introduction of new genes could be carried out by fusion of a protoplast with a liposome or vesicle with entrapped DNA, or fusion of two compatible protoplasts resulting in the combination of two genomes. These are a form of DNA recombination as in fertilization. The hybrid DNA at the active phase of a fused protoplast would be the proper time for microinjection of the foreign genes into the nucleus. Success has been reported in the introduction of genes into mice by microinjection of the fertilized egg (Palmiter et al., 1983). The promoter of the mouse gene for metallothionein-1 was fused to the structural gene coding for the human growth hormone. The mice with incorporated human growth hormone genes grew significantly larger than the control mice. However, expression of the foreign genes in mice is influenced by the site of gene integration and tissue environment.

Foreign gene introduction could also be carried out by microinjection into a nucleus at the active phase of a protoplast, pollen, newly fertilized egg in an ovule, or a cell in an embryonic tissue such as buds. A transformed protoplast could regenerate plantlets with new traits, and modified pollen could be used for pollination. An initial of a primordium for a lateral bud could be the ideal cell for microinjection to produce a sport.

Insertion of bacterial plasmid DNA by Agrobacterium tumefaciens into the genome of many dicotyledon plant cells is a natural process. The bacteria infect the wounded plant tissue and incite crown gall tumors by transferring a specific segment of the Ti (tumor inciting) plasmid DNA and integrating into plant nuclear DNA (Chilton, 1983). The plant cells can express these integrated bacterial genes. A related organism, A. rhizogenes, can transfer genes (Ri) essential for root induction and the genes integrate into plant nuclear DNA. The bacteria incite hairy root disease on many dicotyledon plants (Ream and Gordon, 1982). These systems could permit introduction of desirable genes into plant DNA by using recombinant DNA techniques to insert the desired genes into the plasmids. However, transfer and integration of the Ti or Ri genes has to be excised to prevent tumorigenesis or the gene expression has to be suppressed so that regeneration of the plantlets with desired traits is possible from the transformed cells.

The Ti or Ri plasmid has been used as a vector for the transfer of foreign genes into plants. The integrated genes can express themselves in the plant cells. However, the introduction of genes by microinjection at present has no vector to lead and integrate them into the correct position of the host DNA. The site of integration appears to be random, and expression of the genes is greatly influenced by the site of insertion and their flanking genes in a cellular environment. Therefore, it is essential to search for vectors which will lead the introduced genes to a proper integration, and to understand the genetic components responsible for plant characteristics.

References

Beasley, C.A. 1977. Ovule culture: fundamental and pragmatic research for the cotton industry. In I. Reinert and Y.P.S. Bajaj (eds.) - Applied and Fundamental Aspects of Plant, Cell, Tissue and Organ Culture. Springer-Verlag, New York. p. 160-178.

Boeshore, M.L., I. Lifshitz, M.R. Hansen and S. Izhar. 1983. Novel composition of mitochondrial genomes in Petunia somatic hybrids derived from cytoplasmic male sterile and fertile plants. Mol. Gen. Genet. 190: 459-467.

Bonga, J.M. 1977. Application of tissue culture in forestry. In I. Reinert and Y.P.S. Bajaj - Applied and Fundamental Aspects of Plant, Cell, Tissue and Organ Culture. Springer-Verlag, New York. p. 93-108.

Campos, F.F. and T. Morgan, Jr. 1958. Haploid pepper from a sperm. J. Hered. 49: 134-137.

Carlson, P.S., H.H. Smith and R.D. Dearing. 1972. Parasexual interspecific plant hybridization. Proc. Nat. Acad. Sci., U.S.A. 69: 2292-2294.

Chen, C.H., F.T. Chen, C.F. Chien, C.H. Wang, S.J. Chang, H.E. Hsu, H.H. Ou, Y.T. Ho and T.M. Lu. 1979. A process of obtaining pollen plants of Hevea brasiliensis Muell.-Arg. Sci. Sinica 22: 81-90.

Cheng, W. and R.H. Smeltzer. 1981. Comparison of tissue culture propagules and seedlings of Pinus taeda grown in a greenhouse. For. Prod. Res., Internat. Paper Co. Tech. Note 58. 11 pp.

Chilton, M.-D. 1983. A vector for introducing new genes into plants. Sci. Amer. 248: 51-59.

Chinese Academy of Agricultural Sciences, Inst. Tobacco Res. 1981. A preliminary study on the heredity and vitality of the progenies of tobacco pollen plants. In H. Hu (ed.) - Plant Tissue Culture. Pitman Advanced Pub. Program, Boston. p. 223-225.

Collins, G.B. and J.W. Grosser. 1984. Embryo culture. In I.K. Vasil (ed.) - Cell Culture and Somatic Cell Genetics of Plants. Vol. 1. Laboratory Techniques. Acad. Press, New York.

Collins, G.B. and R.S. Sadasivaiah. 1972. Meiotic analysis of haploid and doubled haploid forms of Nicotiana otophora and N. tabacum. Chromosoma 38: 387-404.

Darlington, C.D. 1937. Recent advances in cytology. 2nd Ed. Blakiston's, Philadelphia.

David, A. and H. David. 1979. Isolation and callus formation from cotyledon protoplasts of pine (Pinus pinaster). Z. Pflanzenphysiol. 94: 173-177.

Doy, C.H., P.M. Gresshoff and B.G. Rolfe. 1973. Time course of phenotypic expression of E. coli gene Z following transgenosis in haploid Lycopersicon esculentum cells. Nature, New Biol. 244: 90-91.

Gengenbach, B.G. 1977. Genotypic influences on in vitro fertilization and kernel development in maize. Crop Sci. 17: 489-492.

Guha, S. and S.C. Maheshwari. 1964. In vitro production of embryos from anthers of Datura. Nature 204: 497.

Guries, R.P. and R.F. Stettler. 1976. Prefertilization barriers to hybridization in the poplars. Silvae Genet. 25: 37-44.

Hagman, M. 1975. Incompatibility in forest trees. Proc. Roy. Soc. London, Ser. B. 188: 313-326.

Hagman, M. and L. Mikkola. 1963. Observations on corss-, self-, and interspecific pollinations in Pinus peuce Griseb. Silvae Genet. 12: 73-79.

Harms, C.T. 1983. Somatic incompatibility in the development of higher plant somatic hybrids. Q. Rev. Bio. 58: 325-353.

Heilungkiang Inst. For., Tree Improv. Lab. 1975. Induction of haploid poplar plants from anther culture in vitro. Sci. Sinica 18: 669-777.

Heinz, D.J. and G.W.P. Mee. 1969. Plant differentiation from callus tissue of Saccharum species. Crop Sci. 9: 346-348.

Heinz, D.J., M. Krishnamurthi, L.G. Nickell and A. Maretzki. 1977. Cell, tissue and organ culture in sugarcane improvement. In I. Reinert and Y.P.S. Bajaj (eds.) - Applied and Fundamental Aspects of Plant, Cell, Tissue and Organ Culture. Springer-Verlag, New York. p. 3-17.

Ho, R.H., Y.A. Raj and L. Zsuffa. 1983. Poplar plants through anther culture. In R.T. Eckert (ed.) - Proc. 28th N.E. Tree Improv. Conf., Durham, New Hampshire, July 7-9, 1983. Northeast. For. Exp. Sta., USDA For. Service. p. 294-300.

Huhtinen, D. 1978. Callus and plantlet regeneration from anther culture of Betula pendula Roth. In T.A. Thorpe (ed). - 4th Int. Cong. Plant Tissue and Cell Culture, Calgary. p. 169.

Illies, Z.M. 1974. Induction of haploid parthenogenesis in aspen by postpollination treatment with toluidine-blue. Silvae Genet. 23: 221-226.

Johnson, C.D., D. Grierson and H. Smith. 1973. Expression of lambda plac5 DNA in cultured cells of higher plants. Nature, New Biol. 244: 105-106.

Johnsson, H. 1953. The early development of hybrid aspen and attempt to predict its future production. Svenska Skogsv. Foren. Tidskr. p. 73-96.

Kanta, K., N.S. Rangaswamy and P. Maheshwari. 1962. Test-tube fertilization in a flowering plant. Nature 194: 1214-1217.

Kao, K.N. 1977. Chromosomal behaviour in somatic hybrids of soybean - Nicotiana glauca. Mol. Gen. Genet. 150: 225-230.

Kasha, K.J. 1974. Haploids from somatic cells. In K.J. Kasha (ed.) - Haploids in Higher Plants, Advances and Potential. Proc. 1st Internat. Symp. Univ. Guelph, Guelph. p. 67-87.

Keller, W.A. and G.R. Stringham. 1978. Production and utilization of microspore-derived haploid plant. In T.A. Thorpe (ed.) - Frontier of Tissue Culture, Calgary, Canada. p. 118-122.

Khoshoo, T.N. 1959. Polyploidy in gymnosperms. Evolution 13: 24-39.

Kleinhofs, A., C.E. Francine, M.-D. Chilton and A.J. Bendich. 1975. On the question of the integration of exogenous bacterial DNA into plant DNA. Pro. Nat. Acad. Sci., U.S.A. 72: 2748-2752.

Kossuth, S.V. and G.H. Fechner. 1973. Incompatibility between Picea pungens Engelm. and Picea engelmanii Parry. For. Sci. 19: 50-60.

Kriebel, H.B. 1972. Embryo development and hybridity barriers in white pines. Silvae Genet. 21: 39-44.

Kriebel, H.B. 1975. Interspecific incompatibility and inviability problems in forest trees. Proc. 14th Can. Tree Improv. Assoc., Part 2, Dept. Environ., Ottawa. p. 67-79.

Krishnamurthi, M. and J. Tlaskal. 1974. Fiji disease resistant Saccharum officinarum var. Pindar subclones from tissue cultures. Proc. Int. Soc. Sugar Cane Technol. 15: 130-137.

Laibach, F. 1925. Das Toubwerden von Bastardsmen und die Kunstaiche Aufzucht fruh absterbender Bastardembryonen. Z. Bot. 17: 417-459.

Larkin, P.J. and W.R. Scowcroft. 1981. Somaclonal variation - a novel source of variability from cell cultures for plant improvement. Theor. Appl. Genet. 60: 197-214.

Lester, D.T. and J.G. Berbee. 1977. Within-clone variation among black poplar trees derived from callus culture. For. Sci. 23: 122-131.

Maheshwari, N. 1958. In vitro culture of excised ovules of Papaver somniferum. Science 127: 342.

Malhotra, K., A. Rashid and S.C. Maheshwari. 1979. Transgenosis in higher plant cells - a reevaluation. Z. Pflanzenphysiol. 95: 21-31.

Melchior, G.H. and F.W. Seitz. 1968. Interspezifische Kreuzungssterilität innerhalb der pappelsektion Aigeiros. Silvae Genet. 17: 88-93.

Mergen, F., J. Burley and G.M. Furnival. 1965. Embryo and seedling development in Picea glauca (Moench) Voss after self-, cross- and wind-pollination. Silvae Genet. 14: 188-194.

Mikkola, L. 1969. Observations on interspecific sterility in Picea. Ann. Bot. Fenn. 6: 285-339.

Nakamura, A., T. Yamada, N. Kadotoni, R. Itagaki and M. Oka. 1974. Studies on the haploid method of breeding in tobacco. SABRAO J. 6: 107-131.

Nishi, T. and S. Mitsuoka. 1969. Occurrence of various ploidy plants from anther and ovary culture of rice plants. Jap. J. Genet. 44: 341-346.

Nitsch, J.P. 1951. Growth and development in vitro of excised ovaries.
Amer. J. Bot. 38: 566-577.

Nitsch, J.P. and C. Nitsch. 1969. Haploid plants from pollen grains.
Science 163: 85-87.

Orr-Ewing, A.L. 1957. A cytological study of the effects of self-pollination on
Pseudotsuga menziesii (Mirb.) Franco. Silvae Genet. 6: 179-185.

Palmiter, R.D., G. Norstedt, R.E. Gelinas, R.E. Hammer and R.L. Brinster. 1983.
Metallothionein-human GH fusion genes stimulate growth of mice. Science 222:
809-814.

Piatnitsky, S.S. 1947. On pollination in oaks and the germination of the pollen on
the stigma. Dokl. Akad. Nauk. USSR 56: 545-547.

Redforth, N.W. and L.C. Pegoraro. 1955. Assessment of early differentiation in
Pinus proembryo transplanted to in vitro conditions. Trans. Roy. Soc. Canada
49: 69-82.

Radojevic, L. 1978. In vitro induction of androgenic plantlets in Aesculus
hippocastanum. Protoplasma 96: 369-374.

Ream, L.W. and M.P. Gordon. 1982. Crown gall disease and prospects for genetic
manipulation of plants. Science 218: 854-859.

Redenbaugh, K., D.F. Karnosky and R.D. Westfall. 1981. Protoplast isolation and
fusion in three Ulmus species. Can. J. Bot. 59: 1436-1443.

Reed, G.M. and G.B. Collins. 1980. Histological evaluation of seed failure in
three Nicotiana interspecific hybrids. Tob. Sci. 24: 154-156.

Shepard, J.F., D. Bidney and E. Shahin. 1980. Potato protoplasts in crop
improvement. Science 28: 17-24.

Stewart, J.M. 1981. In vitro fertilization and embryo rescue. Environ. Exp.
Bot. 21: 301-315.

Stewart, J.M. and C.L. Hsu. 1977. In ovulo embryo culture and seedling development
of cotton (Gossypium hirsutum L.). Planta 137: 113-117.

Tanaka, M. and K. Nakata. 1969. Tobacco plants obtained by anther culture and the
experiment to get diploid seeds from haploids. Jap. J. Genet. 44: 47-54.

Thomas, M.J. 1972. Compartement des embryons de trois especes de pins, isoles au
moment de leur elivage et cultives in vitro, en presence de
cultures-neurrices. C.R. Acad. Sci. Paris 274: 2655-2658.

Wang, C.C., Z.C. Chu and C.S. Sun. 1975. The induction of Populus pollen-plants.
Bot. Sinica 17: 56-59.

Wenzel, G. 1980. Anther culture and its role in plant breeding. In Symp. on plant
tissue culture, genetic manipulation and somatic hybridization of plant
cells. Bhabha Atomic Res. Centre, Bombay.

Withner, C.L. 1942. Nutrition experiments with orchid seedlings.
    Amer. Orchid Soc. Bull. 11: 112-114.

Zenkteler, M. 1980. Intraovarian and in vitro pollination. In I.K. Vasil (ed.) -
    Perspectives in Plant Cell and Tissue Culture. Internat. Rev. Cytol. Suppl.
    11B. Acad. Press, New York. p. 137-156.

Zhu, X.Y., R.L. Wang and Y. Liang. 1980. Induction of poplar pollen plants.
    Sci. Silvae Sinicae 16(3): 190-197.

Steam Pretreatment of Aspenwood for
Enhanced Enzymatic Hydrolysis

H.H. Brownell, M. Mes-Hartree and J.N. Saddler
Forintek Canada Corp.
800 Montreal Road, Ottawa, Canada K1G 3Z5

ABSTRACT

The effect of moisture content (MC) on the rate of temperature rise inside simulated wood chips, while heated in saturated steam at 240°C, has been measured. Rapid heating inside dry chips results from fast steam penetration and condensation inside the wood pores. Available pore volume limits the amount of condensation inside chips of MC > 100% (OD basis), necessitating slow heat transfer by conduction, causing uneven cooking and poorer results with high MC chips. Solubilization of pentosan, and yield of glucose on subsequent enzymatic hydrolysis, were only marginally lower from air-dry than from green chips. After treatment with 3.24 MPa steam at 240°C for 3 minutes, bleed-down of pressure to only 0.34 MPa before explosion, resulted in similar yields of glucose and of total reducing sugars, and similar yields of butanediol plus ethanol on fermentation, as were obtained without bleed-down. Analysis of literature data indicates high temperatures to be unnecessary, other than to reduce treatment time.

INTRODUCTION

The treatment of wood by high pressure saturated steam at temperatures of up to 285°C has long been practised in the Masonite Process (1) for the production of fibreboard. In this process, high temperatures soften the wood, and mechanical action during high pressure discharge, results in the necessary fibre separation. Chemical hydrolysis is also involved, as is shown by the solubilization of much of the hemicellulose, which is sold as "wood molasses" for use as a supplement to cattle feed. DeLong (2,3) in a similar steam explosion process, obtained an exploded product suitable for ruminant fodder, and as a substrate for enzymatic hydrolysis. DeLong emphasized the importance of the mechanical effect and of raising the temperature above the softening points of the lignin, hemicellulose and cellulose components. Foody (4), using a similar steam gun, recognized the importance of the chemical processes involved, and indicated a wide range of temperatures to be satisfactory, provided the time of treatment was correspondingly adjusted. Marchessault (5), working with DeLong's product, showed that it could be separated into 3 streams by extraction with water, which removed hemicellulose, and then with dilute alkali which removed much of the lignin. Taylor (6) described continuous steam-treatment equipment and the "Stake Biomass Conversion Process" which included a similar separation into 3 streams. Gast et al (7) described an organosolv separation into 3 streams without explosion. Puri and Mamers (8), however, reported data indicating increased enzymatic digestibility resulting from mechanical action during discharge from their "Siropulper".

In the present paper, the effects of the magnitude of the pressure drop during steam explosion, and of the initial moisture content of the wood put into the steam gun, are described. Published data on the effect of steam temperature are also assessed.

METHODS

Steam gun: The steam gun was an insulated vertically mounted stainless steel cylinder of inside diameter 63 mm and volume 2 litres, closed at the lower end by a stainless steel Kamyr ball valve, and at the upper end by a Parr bomb lid. The lid was fitted with a pressure gauge, a bleed valve, and thermocouple probes extending down into the gun. Steam, from a high pressure boiler, was introduced through the side of the barrel, immediately below the lid. The gun was preheated by pressurization with steam at the desired temperature. Wood chips, of equivalent dry weight 200 g, were loaded into the empty preheated gun and, after a steam treatment, the contents of the gun were explosively discharged through the ball valve into a stainless steel cyclone, from which the wet exploded wood was recovered. In some experiments, after the steam treatment, the steam pressure was bled down to 0.34 MPa from the treatment pressure of 3.24 MPa, before explosion to atmospheric pressure.

Wood: The wood used was from logs of freshly cut and peeled green Populus tremuloides, stored frozen until required. Chips were 3.3 mm in thickness, with the fibre direction at an angle of about 45 degrees to the chip face. The moisture content of the green chips was 51.97% (moisture-containing basis), and the pentosan content was 18.35% (OD basis).

Temperature rise within wood specimens: Thermocouple probes were inserted tightly to the bottom of holes drilled axially in small sapwood specimens of Populus tremuloides, as shown in Fig. 1. The top of the annular space, between probe and specimen, was sealed with fast curing epoxy resin, forced in by means of a tight-fitting bushing pressed down into a shallow hole of the same diameter in the top of the speciment. End coating, or coating of the entire specimen, where indicated, was with a thin layer of fast-curing epoxy resin.

Water extraction: Water washing of the steam exploded wood to remove soluble pentosan was done by vigorous mechanical stirring of the material as a 5% slurry for 2 hours in distilled water. After recovery by vacuum filtration, the washed material was washed twice more (1 hour each), also at 5 percent consistencies, with recovery by filtration.

Analyses: Pentosan analyses were done by TAPPI Standard Method T 223-os-71. Glucose was determined by the glucostat enzyme assay of Raabo and Terkildsen (9), and total reducing sugars by the colorimetric dinitrosalicylic acid method of Miller (10).

Enzymatic hydrolysis: Enzymatic hydrolysis of washed steam-exploded wood, or or Solka Floc, was with a culture filtrate of Trichoderma harzianum E58, at a substrate consistency of 5%, for a period of 24 hours.

Combined hydrolysis and fermentation: The method used for combined enzymatic hydrolysis and fermentation with Klebsiella pneumoniae, has been described by Yu et al (11).

RESULTS AND DISCUSSION

The rates of increase in temperature inside aspenwood specimens heated in saturated steam at 240℃ are shown in Fig. 2, 3 and 4 . Fig. 2 shows that

38

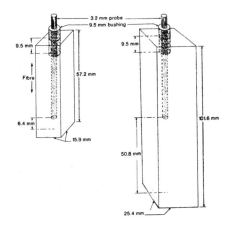

Fig. 1   Thermocouple probes inserted in dowels of square cross section, with
the upper end of the annular space between the probe and dowel
sealed with a bushing

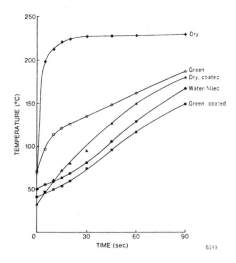

Fig. 2   Effect of moisture content of aspenwood dowels (15.9 x 57.2 mm),
and of sealing entire outer surface of dowels with epoxy resin, on
rate of temperature rise inside the dowels, 6.4 mm from lower end (as
measured by 3.2 mm thermocouple probe), when heated in 2L steam gun
with saturated steam at 240°C.

the temperature of the sensitive tip of a thermocouple probe, 3.2 mm in diameter, surrounded by air-dry aspen sapwood of thickness 6.4 mm, rose very rapidly when suspended in saturated steam at 240°C. The temperature reached 200°C, for example within 5 seconds of opening the steam valve. Comparable specimens of green aspen sapwood (approximately half wood substance and half water) rose much more slowly in temperature, requiring more than 1.5 minutes to reach 200°C. Successively slower rates were obtained for a dry specimen coated with a thin film of epoxy resin, for a water-filled specimen, and for a green specimen coated with epoxy resin. The steam apparently rapidly penetrated the air-dry uncoated specimen and, by condensing inside the void volume, supplied its latent heat within the specimen. Dry coated specimens or water-filled specimens, could not be penetrated and were consequently heated by the slower process of conduction. Also, more heat was required to heat the green and the water-filled wood, because the heat capacity of water is approximately double that of dry wood substance.

Fig. 3 shows that the coating of only the end grain of an air-dry specimen of aspen sapwood, retarded the heating by a substantial amount, when compared with the effect produced by coating the entire outer surface (sides as well as ends). Entry of steam was, as expected, mainly through the end grain. Furthermore, when the much larger air-dry specimen shown in Fig. 4 was used, with the thickness of wood surrounding the sides of the probe being almost doubled, and the sensitive tip of the probe located 50.8 rather than 6.4 mm from the end of the specimen, the temperature rise of the probe tip was equally fast. End coating of the large air-dry specimen resulted in the slowest rate of temperature rise observed in all the experiments. Clearly, when end grain of air-dry aspen sapwood was available, the steam entered and passed throughout the specimen within a few seconds, over distances of at least 50 mm in the fibre direction.

When aspenwood chips are heated from room temperature up to steam temperature with saturated steam, the theoretical amount of steam, which must be condensed to supply the required heat, increases with the initial moisture content of the chips. If this steam condenses inside the pores of the wood, the theoretical final moisture content of the heated chips should correspondingly increase, as is exemplified by the two solid lines in Fig. 5 for steam temperatures of 250 and 210°C. In practice, however, the pore volume is limited, and as can be seen from the position of the dotted line, if the initial moisture content is 49 percent (moisture-containing basis), the pore volume is theoretically filled with water by the time the steam temperature of 250°C is reached.

Aspenwood chips of this porosity, and of initially higher moisture content, cannot be heated all the way to this steam temperature by the rapid mechanism of condensation inside the wood. The much slower heating of wetter chips by conduction should, therefore, result in a poorer product with over-cooked exteriors or under-cooked interiors. Interestingly, an examination of the data reported by Foody (12) shows that the maximum values of in vitro cellulose digestibility (IVCD) and also the maximum yields of both glucose and of xylose on enzymatic hydrolysis, were obtained with chips of about 35 percent initial moisture content (moisture-containing basis). When the initial moisture content was increased to about 50 percent, the IVCD, did, in fact, decrease by about 1 part in 10 or 20, at each of four different steam temperatures ranging from 208 to 250°C. Similar decreases of 1 part in 10 in the glucose yield, and of 1 part in 3 or 4 in the more sensitive xylose yield, were also obtained, consistent with less uniformly cooked products.

40

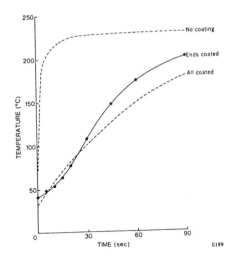

Fig. 3 Rate of temperature rise inside air-dry aspenwood dowel
(15.9 x 57.2 mm) 6.4 mm from lower end when upper and lower
ends, only, were coated with epoxy resin (solid line), compared
with rates for uncoated and for completely coated (sides as well
as ends) dowels (dotted lines), heated in saturated steam at 240°C.

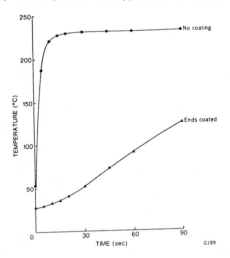

Fig. 4 Rate of temperature rise in centre (50.8 mm from ends) of large
air-dry aspenwood dowel (25.4 x 106 mm) and comparison with
rate when the ends, only, of the dowel were coated with epoxy
resin, when heated in saturated steam at 240°C

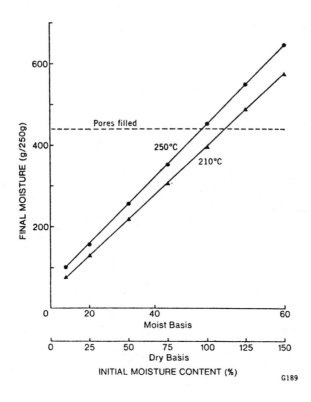

Fig. 5   Effect of steam temperature, and of initial moisture content of
aspenwood chips, on theoretical final moisture content (initial
moisture content plus condensed steam) of heated chips at steam
temperature, and comparison of this moisture content with that at
which pores are completely filled with water, marking end of
heating by steam penetration

Foody's data also show that, when the initial moisture content of wood
chips was reduced below the optimum of 35 percent a poorer product was
obtained at each of 5 different steam temperatures. By air-drying to
about 6 or 8 percent MC (moisture-containing basis), the IVCD value was
reduced generally by approximately 1 part in 5, and the glucose yield
by 1 part in 5 or 10.   These results were obtained despite the fact that
rapid and uniform heating of the air-dry chips should have occurred, as
indicated by Fig. 2. Another effect of reducing the moisture content
from 52 percent (green wood) to 6 percent (air-dry wood), was to decrease
the steaming time required for both optimum glucose yield and for maximum
IVCD, by an average of 40 or 50 percent, at 4 different steam temperatures.
These shorter optimum times for drier chips, reflect the faster heating
shown in Fig. 2.

In the present study, using smaller aspenwood chips, the effect of initial moisture content on the pentosan content of the exploded wood product, and on the yields of glucose and of total reducing sugars obtained on enzymatic hydrolysis, are shown in Table 1. Solubilization of the pentosan fraction of the air-dry chips was marginally lower than that of the green chips steamed for the same time, despite the faster heating of the former. The yields of glucose and of total reducing sugars from the air-dry chips were also marginally lower. A longer steaming time (180 vs 80 seconds) increased the yields, showing that the optimum time had not been passed. These data show that small air-dry chips could be used with little, if any, sacrifice in sugar yield. The lower degree of hydration of the air-dry chips may have reduced the susceptibility to hydrolysis during steam treatment, overcoming the expected accelerating effect of the more rapid heating.

Table 1. Effect of moisture content of aspenwood chips (3.5 mm thick in fibre direction) heated in saturated steam at 221°C, on solubilization and destruction of pentosan, and on yield of total reducing sugars and of glucose on subsequent enzymatic hydrolysis of water-washed exploded wood

Moisture content % (wet basis)	Time of steaming (s)	Pentosan analysis[1]		Enzymatic hydrolysis[2] of washed exploded wood	
		exploded wood %	water-washed exploded wood %	reducing sugars[3] mg/mL	glucose[4] mg/mL
green 51.97	80	14.97	6.22	5.2	4.8
air-dry 6.68	80	15.30	8.06	4.7	4.1
air-dry 6.68	180	9.69	3.25	5.9	5.8
Solka Floc(control)	0	-	-	16.3	12.6

Footnotes

[1] Pentosan analysis of original aspenwood: 18.35%

[2] 24 hours at a 5% consistency (of washed exploded wood or of Solka Floc) with culture filtrate of T. harzianum E58

[3] Reducing sugars by DNS

[4] Glucose by glucostat

The steam-explosion process is characterized by high steam temperatures for short times, followed by explosive decompression (2). The necessity, however, for either the high temperature, or the explosion, is questionable, when the process is directed to the pretreatment of biomass for enzymatic hydrolysis or microbiological utilization. Examination of Foody's (12) data shows that the maximum values of IVCD (under the best conditions of initial moisture content in each case) actually decreased regularly, with increasing steam temperature, through 5 steam temperatures ranging from 208 to 259°C. The decrease apparently resulted from a relatively greater formation of toxic decomposition products at the higher temperatures. In the case of enzymatic hydrolysis, where living organisms were not involved, the maximum yields of glucose, obtained after treatment with steam at 226 and 250℃, differed by an insignificant 2 parts in 100. No advantage in the use of the higher temperature was therefore apparent, at least for cellulose hydrolysis.

In the present work, the effect of the magnitude of the pressure drop during explosive decompression, after steam treatment for 180 seconds at 240℃, was examined. When the pressure was bled-down from the treatment pressure of 3.24 MPa to only 0.34 MPa (3.4 atmospheres), before explosion to atmospheric pressure from this reduced pressure, the yields of glucose and of total reducing sugars, obtained on subsequent enzymatic hydrolysis, were both reduced by an apparently insignificant 1 part in 18. The corresponding yields of butanediol plus ethanol, obtained on combined enzymatic hydrolysis and fermentation with Klebsiella pneumoniae, were unchanged within the accuracy of the data, the apparent reduction being only 1 part in 70. The mechanical effect of the explosion (i.e. the "big bang") was not necessary.

One feature of the steam explosion process (2,3) is the separation of wood, or other biomass, into three streams (cellulose, hemicellulose, and lignin), the hemicellulose fraction becoming soluble in water, and the lignin in ethanol or in alkali. Separation of the hemicellulose fraction, however, is practised commercially in the pulping industry at much lower temperatures and pressures. In such a prehydrolysis cook, wood chips are heated in water with or without added acid. The amounts of pentosan solubilized, and recovered from solution after steam explosion at 240 and 220℃, are compared in Fig. 6 with the corresponding solubilization and recovery obtained by Springer and Harris (13) by prehydrolysis at the much lower temperature of 170℃. Fig. 6 shows no apparent advantage of the high temperatures of steam explosion as far as pentosan solubilization is concerned. In both cases, the addition of acid increased the desired solubilization over the undesired destruction. The treatment time required at 170C is, of course, much longer than that at 240℃, but the steam pressure is only one-fifth as great, permitting the use of larger equipment. Pentosan solubilization, of course, is not the only factor influencing accessibility. Possible adverse effects on the lignin, under different treatment conditions, must also be considered.

CONCLUSIONS

1) Air-dry aspen sapwood chips heat much faster in the steam gun then do green chips. Fast heating, by steam condensation inside the wood pores, ceases when the pores become filled with water. Slow heating of wetter wood is by conduction, and gives less uniform cooking.

2) Optimum reported yields of sugars, on enzymatic hydrolysis of steam exploded wood, occur at about 36 percent original MC (moisture-containing basis). Higher MC gives less-uniform heating. Lower MC wood is apparently somewhat less reactive to steam hydrolysis.

3) Optimum steaming times increase markedly with increased initial MC.

4) Optimum yields of glucose, on subsequent enzymatic hydrolysis, vary insignificantly with steam temperature below 250℃.

44

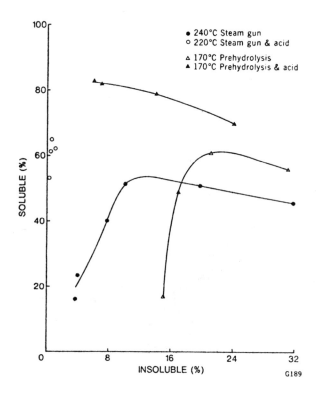

Fig. 6   Percent of original pentosan of aspenwood solubilized as a result
of steam-explosion treatment, with and without impregnation with a
0.2 percent solution of sulfuric acid, compared with analogous
solubilization at lower temperature, during prehydrolysis treatment
with water and with a 0.4 percent solution of sulfuric acid, all shown
as functions of the amount of original pentosan still remaining
undissolved (Prehydrolysis data adapted from Springer and Harris)

5) The mechanical effect of explosion from high pressure is unnecessary,
when the wood has received an adequate cook, and is intended for enzymatic
hydrolysis or combined hydrolysis and fermentation.

6) High steam temperatures (such as 240 or 220°C) are unnecessary for the
solubilization of the hemicellulose fraction. Equally good, or better,
results can be obtained at 170°C.

REFERENCES

1) Koran, Z., B.V. Kokta, J.L. Valade and K.N. Law, 1978. Fibre Characteristics of Masonite Pulp. Pulp & Paper Canada 79 (3): T107-T113.

2) DeLong, E.A. 1981. Method of Rendering Lignin Separable from Cellulose and Hemicellulose in Lignocellulosic Material and the Product so Produced. Canadian Patent No. 1096374 (Feb. 24)

3) DeLong, E.A. 1983. Method of Rendering Lignin Separable from Cellulose and Hemicellulose and the Product so Produced. Canadian Patent No. 1141376 (Feb. 15).

4) Foody, P. 1984. Method for Obtaining Superior Yields of Accessible Cellulose and Hemicellulose from Lignocellulosic Materials. Canadian Patent 1163058.

5) Marchessault, R.H., S. Coulombe, T. Hanai and H. Morikawa. 1980. Pulp Paper Mag. Can. Trans. 6: TR52-TR56.

6) Taylor, J.D. 1980. Continuous Autohydrolysis, A Key Step in the Economic Conversion of Forest and Crop Residues into Ethanol. Energy from Biomass. Proc. of Int. Conf. Biomass, Brighton, England. pp. 330-336 Nov. 4-7.

7) Gast, D., C. Ayla and J. Puls. 1982. Component Separation of Lignocelluloses by Organosolv Treatment. EC Conference, Energy from Biomass. Berlin Sept. 20-23, Appl. Science Publishers Ltd. London, 1983. pp. 879-881.

8) Puri, V.P. and H. Mamers. 1983. Explosive Pretreatment of Lignocellulosic Residues with High-Pressure Carbon Dioxide for the Production of Fermentation Substrates. Biotechnol. Bioeng. 25: 3149-3161.

9) Raabo, E. and T.C. Terkildsen. 1960. On the Determination of Blood Glucose, Scand. J. Clin. Lab. Invest. 12: 402-406.

10) Miller, G.L. 1959. Use of Dinitrosalicylic Reagent for the Determination of Reducing Sugars. Anal. Chem. 31: 426-428.

11) Yu, E.K.C. 1984. The Combined Enzymatic Hydrolysis and Fermentation of Hemicellulose to 2,3-Butanediol. Appl. Microbiol. Biotechnol 19: 365-372.

12) Foody, P. 1980. Optimization of Steam Explosion Pretreatment. Final report to the U.S. DOE, Contract AC0279ETZ3050. April.

13) Springer, E.L. and J.F. Harris. 1982. Prehydrolysis of Aspenwood with Water and with Dilute Aqueous Sulfuric Acid. Svensk papperstidning 1982. 152-154.

PRETREATMENT OF LIGNOCELLULOSICS FOR BIOCONVERSION APPLICATIONS: PROCESS OPTIONS

JONATHAN LAMPTEY, MURRAY MOO-YOUNG AND CAMPBELL W. ROBINSON

Process Biotechnology Research Group,
Department of Chemical Engineering,
University of Waterloo, Ontario, Canada N2L 3G1

ABSTRACT

The potential for using lignocellulosic biomass materials in bioconversion processes is well recognized. A major problem in the commercialization of this potential is the inherent recalcitrance of these materials to biological transformations. If the production rate and yield of fermentable glucose sugar from lignocellulosics could be increased economically beyond currently attainable levels, their use for the manufacture of various fermentation products (e.g., antibiotics, alcohols, solvents, organic acids, etc.) would become more attractive. Unfortunately, the available processes usually require severe chemical treatment (resulting in low yields as well as toxic and/or inhibitory by-products) and/or high energy inputs.

Depending on the nature of the eventual feedstock application, various processes have been proposed for the pretreatment of lignocellulosic materials. The rationale for these processes (e.g., mechanical shearing, steam explosion, organosolv and acid), some involving several steps, is reviewed in terms of their effectiveness (in the fractionation and/or hydrolysis of the biomass) and the need to minimize the formation of sugar decomposition products. Case studies from our experience relate to: autohydrolysis, combined alkali-gamma irradiation and acid pretreatment; solvent delignification; and enzymatic hydrolysis. The test materials used were corn stover, hybrid poplar, and Kraft pulpmill sludge. For the corn stover and pulpmill sludge cases, a brief comparison of economic sensitivity analyses of the best available technologies for the production of fuel ethanol, methane biogas and feed SCP indicates the relative unattractiveness of ethanol production at the present time.

INTRODUCTION

In a world of diminishing resources and increasing needs, every opportunity for the reuse of waste materials must be examined. Waste lignocellulosic materials from agriculture, agro-industrial processing and forestry can help fulfill the requirements for inexpensive food, fuel, fertilizer and various chemicals. Rising petroleum prices, foreign exchange imbalances and pollution-related problems impel research into the economical exploitation of lignocellulosic materials. Caution should temper optimism, however, when initiating waste biomass utilization projects. In each case, the alternatives must be carefully evaluated to assess their economic potential.

Current research efforts to develop an economically viable technology for the utilization of cellulosic biomass materials, in Canada and elsewhere, are based on the availability of abundant quantities of agricultural and forestry residues that presently have little or no economic value. Canada, for example, produces about 14×10^7 m^3 of wood residues and by-products per year which could be used for the production of valuable products (Phillips et al., 1979). The fact that there are virtually no commercial processes based on the large-scale use of lignocellulosics reflects the uneconomic state of known technologies to produce glucose sugar, the basic raw material for the manufacture of fermentation products. Currently proposed technologies suffer from unacceptably low yields and/or rates of production of truly fermentable sugars from the real raw materials. Among the various reasons for this unfortunate situation are the following technical constraints in using native lignocellulosics (exemplified by wood) as direct feedstocks: (1) difficulty of materials handling the solids in a continuous processing system, (2) formation of undesirable degradation products in known chemical conversion processes, (3) substrate recalcitrance in known bioconversion processes. Structural features of cellulose, such as the degree of crystallinity, and the degree of polymerization, limit accessibility of the cellulose substrate to acid and enzyme action.

There are two main techniques for producing sugar solutions from the cellulosic components of lignocellulosic biomass: (1) acid hydrolysis, and (2) enzymatic hydrolysis. The required high temperatures and pressures for acid hydrolysis result in significant product degradation and, hence, low yields. With regard to enzymatic hydrolysis, although the yields are better, the conversion rates are low and the enzyme ingredient is expensive. In order to improve the efficiency of these two methods, an additional front-end pretreatment step must be incorporated in the processing scheme.

In this paper, the rationale for the variously proposed pretreatment techniques are reviewed in terms of their effectiveness, and the need to minimize the formation of sugar decomposition products. A brief discussion of case studies from our experience -- autohydrolysis, combined alkali-gamma irradiation, alkali, and acid pretreatment; solvent delignification and enzymatic hydrolysis -- is also presented.

BACKGROUND: REVIEW OF PRETREATMENT TECHNIQUES

Perhaps the most important economic consideration is that of substrate pretreatment. Pretreatment costs are the second most important expense after raw materials costs in any saccharification process (Ladisch, 1979; Lamptey, ' 1983). These costs include not only reagent costs, but also capital, labour and overhead. Most of the proposed pretreatment techniques are too expensive. For others, the subsequent hydrolysis times are generally too long or the conversions, although improved over the untreated material, are not nearly complete.

In order to be effective, pretreatment techniques for enhancing the chemical and enzymatic reactivity of cellulosic materials must alleviate two major constraints: the lignin seal, which restricts enzymatic and microbiological access to the cellulose; and cellulose crystallinity, which limits the rate of all forms of attack on the cellulose. Thus, in developing more effective pretreatment techniques for natural cellulosic materials, particular emphasis should be given to the physical and chemical methods by which crystallinity and the lignin barrier can be overcome. Particular attention must be directed at techniques whose major action is that of crystallinity reduction with a concomitant increase in the rate of sugar formation and a decrease in the rate of formation of sugar decomposition products. While the intracrystalline swelling techniques of mercerization, regeneration and amine de-crystallization do enhance hydrolytic action, their economic feasibility is

questionable. By far the most successful techniques for effecting a drastic alteration of cellulose crystallinity are the two physical pretreatments: vibratory ball
milling and electron irradiation. Both provide dramatic increases in the rates of
hydrolysis and in the yields of sugar under dilute acid saccharification conditions.
The shearing and compressive forces involved in milling cause a reduction in
crystallinity, a decrease in the mean degree of polymerization, an increase in surface area, and an increase in bulk density. The increase in bulk density allows for
a high slurry concentration, thereby reducing the reactor volume and, hence, capital cost (Fan et al., 1982).

The electron irradiation and grinding/milling techniques demand expensive
energy, and their practical adaptation hinges on favourable results from further
research and development. In order to reduce the energy requirements in the
irradiation process, the combined action of swelling agents (mainly NaOH) and
gamma radiation (from nuclear wastes, for example) have been investigated.
Although, by the combined treatment, the lignocellulosic materials were extensively
degraded and most of the solid solubilized, the product was found to contain little
glucose. Most of the soluble carbohydrates appeared to be cellodextrin and
smaller molecular weight glucose degradation products. Any glucose produced by
gamma radiolysis was undoubtedly decomposed by the presence of alkali and the
high dosage of gamma radiation (usually greater than 120 Mrad). Hence, the direct
production of glucose by the combined treatment is difficult. As shown in this
report, a more reasonable approach is a mild irradiation as a pretreatment followed
by acid or enzymatic hydrolysis.

Another technique of current interest, which involves both physical and chemical aspects, is the steam-explosion process (Muzzy et al., 1982). With this technique, lignocellulosic materials (usually wood chips) are saturated with water under
pressure (300–500 psig) at elevated temperatures (215–260°C). When the pressure
is released, the water evaporates rapidly and the wood fibers tend to separate,
increasing the surface area for the subsequent hydrolysis. In addition, the moisture and high temperatures involved result in the liberation of acetic acid which
catalyzes the hydrolysis of the hemicellulose fraction. The physical effect of the
process is similar to what occurs when the moisture in a popcorn kernel evaporates violently and the kernel expands markedly, thus producing popcorn. Although
the technique is highly effective, it requires considerable thermal energy in the
form of steam. It has the additional disadvantage that some of the sugars are
inevitably degraded by the high temperatures involved, even at small reaction
times.

Chemical treatment with strong acids or bases, such as sulfuric acid or sodium
hydroxide, also effectively increase the hydrolysis of cellulose (Fan et al., 1982).
These chemicals are generally quite corrosive and expensive, and must be recovered for reuse. In addition, they are often toxic or inhibitory to microorganisms (or
their enzymes) so that their removal from the pretreated cellulosic material must
be almost complete. These factors combine to increase the expense and difficulty
of such chemical treatment methods.

It is important to note that effective utilization of all three major components
of lignocellulosic biomass (cellulose, hemicellulose, lignin) is necessary in order to
have an economically attractive process (Humphrey et al., 1979). The hemicellulose fraction is considerably easier to hydrolyze to pentoses than is cellulose to
hexoses. The selective removal of hemicellulose from the lignocellulosic matrix
could increase the value of the biomass because the residue remaining after hemicellulose removal is more susceptible to subsequent lignin solvent extraction and/
or enzymatic hydrolysis. In addition, in large scale commercial application, the
removal of lignin may result in lower capital and operating costs since significantly

less biomass (15–40%) will enter the enzymatic hydrolysis unit. Consequently, the volume of the hydrolyzer required will be less than if lignin were to be processed. It is not clear whether the reduced capital and operating costs will be offset by solvent recovery costs.

The separation of the lignocellulosic complex into its constituent components (hemicellulose, cellulose, and lignin) may also result in a more optimum and more flexible downstream processing. For example, in ethanol production different processing schemes and different microorganisms are required to ferment the pentose and hexose sugars.

CASE STUDIES

1. Acid Pretreatment

Lee et al. (1983, 1984) presented detailed results on the H_2SO_4 prehydrolysis of corn stover and hybrid poplar. For a 10% w/v slurry of hybrid poplar, Lee et al. (1984) found that the yield of solids left after pretreatment at 120°C for 0.5h was negligible for H_2SO_4 concentrations less than 1.0% w/v. Less than 20% solubilization of the wood was achieved under these conditions. Significant solubilization (30–50%) was achieved in the temperature range 140 to 180°C and for H_2SO_4 concentration of at least 0.5% w/v for a holding time of 1h. From consideration of the extent of solubilization and the need to minimize sugar degradation, the following conditions were found to be the best for the prehydrolysis of hybrid poplar: 140°C, 1h, 0.5% w/v H_2SO_4, 10% w/v slurry. Under these conditions, 4.2g reducing sugar per 100g hybrid poplar were obtained after enzymatic hydrolysis. The total sugar (prehydrolysis and enzymatic hydrolysis) obtained was 28.7g per 100g hybrid poplar. This level of conversion is rather very low.

Lee et al. (1983) also presented results for the H_2SO_4 pretreatment of corn stover. The best conditions were found to be: 2% w/v H_2SO_4, 95°C, 1h, 10% w/v slurry. The prehydrolysis temperature for corn stover is significantly lower than that commonly used for wood. The optimum temperature for hybrid poplar was found to be 140°C. For hardwoods, the approximate conditions suggested for prehydrolysis prior to Kraft pulping are 0.3 to 0.5% w/v H_2SO_4 at 140°C, or water at 160 to 170°C, for 0.5 to 3h process time. For both corn stover and hybrid poplar, Lee et al. (1983, 1984) found that the use of temperatures greater than 95°C and 140°C, respectively, resulted in significant decomposition of the sugars giving a lower overall sugar yield.

2. Gamma-Ray Pretreatment

Moo-Young et al. (1985a) have reported on the effectiveness of combined NaOH-gamma irradiation treatment of pulp mill sludge. They found that the effectiveness of the pretreatment increased with both gamma-ray dosage and NaOH concentration. At a given dosage, the increase in the extent of solubilization was more marked with increasing NaOH concentration. However, the cellulose content of the residual solids decreased more significantly with increasing gamma-ray dosage. A high degree of solubilization was obtained (up to 92 wt%) at gamma-ray doses in the range of 48 to 150 Mrad. At higher irradiation doses (at least 100 Mrad) and at lower doses with high caustic concentration (12.8 wt%), almost complete solubilization was achieved. At all gamma-ray doses with 12.8 wt% NaOH, the cellulose content of the residual solid was 5 wt% or less. It was found that the high solubilization achieved was not reflected in the carbohydrates and reducing sugars produced due to the decomposition of the sugars by the treatment. The degradation of soluble carbohydrates and reducing sugars produced by the combined alkali–gamma irradiation pretreatment method was especially marked on prolonged exposure to the treatment condition.

Similar results were obtained by Gonzales-Valdes et al. (1981) for the combined NaOH–irradiation pretreatment of corn stover (CS). A solubilization of up to 94% of solid material was obtained under the following conditions: dosage, 150 Mrad, NaOH concentration, 0.51g NaOH/g corn stover. In the absence of NaOH, the extent of solubilization increased by a factor of about 2.6 for a dosage of 100 Mrad over the control without gamma irradiation. The extent of solubilization increased from 26.1 wt% to 87.5 wt% as the concentration of NaOH was increased to 0.51g NaOH/g CS at a dose of 100 Mrad. In this range, the total sugars in the hydrolysate increased from 13.8 wt% to a maximum of 32.3 wt% at 0.06 NaOH/g CS and then decreased to 23.5 wt% at 0.51g NaOH/g CS. In spite of the significant solubilization (87.5 wt%) of the corn stover at 100 Mrad and 0.51g NaOH/g CS, the reducing sugar in the hydrolysate was about the same as that obtained in the absence of NaOH. The same conclusions were observed from analysis of the solid residue. There was extensive solubilization of the lignocellulosic complex (cellulose (40–86%) and lignin (40–90%)) at high NaOH concentrations. Extensive solubilization of the cellulose component was not reflected in a high reducing sugar concentration in the hydrolysate.

Gonzales-Valdes et al. (1981) carried out a series of experiments to determine the suitability of the combined NaOH–gamma irradiation technique as a pretreatment scheme for the enhancement of soluble sugar production from corn stover by enzymatic hydrolysis. The NaOH levels and radiation dosages were varied from 0 to 0.51g NaOH/g CS and from 48 to 150 Mrad, respectively. The combined residue and solubles were then hydrolyzed enzymatically at 40°C for 48h. Table 1 shows the yield of reducing sugar after 48h of enzymic saccharification. The overall yield of reducing sugar increased from 16.5% to a maximum of 67.9% as the irradiation dose was increased to 150 Mrad in the absence of NaOH. The effectiveness of the combined treatment on enzymic saccharification increased with increasing NaOH concentration and irradiation dosage up to 100 Mrad after which the effectiveness decreased due to the significant formation of decomposition products. The maximum yield of sugar of 94.2% based on the total carbohydrate content of the original corn stover was obtained at 100Mrad and 0.06g NaOH/g corn stover. At a given irradiation dose, degradation of the solubles formed after the pretreatment was especially significant at NaOH concentrations greater than 0.10g NaOH/g CS.

Although the combined NaOH–gamma irradiation pretreatment is very effective, it is important to realize that optimum sugar production requires careful control of the level of gamma irradiation and NaOH concentration. The technique is not selective in the solubilization of the various components of the lignocellulosic complex. The hydrolysate obtained contains monomers and oligomers of cellulose, hemicellulose and lignins.

3. Solvent Delignification Pretreatment

Lee et al. (1983, 1984) presented detailed results on the pretreatment of hybrid poplar and corn stover with various solvents. The processing schemes examined included delignification of untreated, or of acid prehydrolyzed biomass. Hybrid poplar was initially prehydrolyzed using H_2SO_4 under the following optimal conditions: 10% w/v slurry, 140°C, 1h, 0.5% w/v H_2SO_4. The amount of solid residue obtained after the prehydrolysis was 72.5g solid residue per 100g hybrid poplar and a reducing sugar yield of 24.5g reducing sugar per 100g hybrid poplar. The H_2SO_4 prehydrolyzed hybrid poplar solids were subjected to solvent extraction

with various amines, aliphatic and aromatic alcohols, ethyleneglycols and its derivatives, and with acetic acid. For the various amines (ethanolamine, n-butylamine
(nBA), ethylenediamine) and acetic acid used, the total sugar produced (from prehydrolysis and enzymatic hydrolysis) was not significantly different from that of the
control (which was not subjected to solvent extraction) value of 28.7 g per 100g
hybrid poplar. Less than about 30% of the total sugar produced was obtained
from the enzymatic hydrolysis stage. This was even true for runs in which over
40% of the prehydrolyzed solids were solubilized. The almost complete removal of
lignin was not reflected in a quantitative conversion of the cellulose to glucose. It
is possible that some degree of recrystallization of the cellulose is associated with
delignification of the prehydrolyzed solids.

Analysis of the acid prehydrolyzed, solvent-extracted hybrid poplar solid residues, for cellulose and lignin showed extensive delignification (over 50%) with
n-butylamine (at 170°C), and with ethylcellosolve (almost complete delignification)
at 170°C. In spite of the almost complete delignification with ethylcellosolve at
170°C, the reducing sugar produced after enzymatic hydrolysis was only marginally higher than that obtained from the control without delignification. It is not
clear whether this is due to an increase in recalcitrance of the cellulose or to possible toxicity from traces of solvent in the hydrolysis mixture. The results for
enzymatic hydrolysis of solvent delignified hybrid poplar solids which had not initially been prehydrolyzed showed that no significant delignification was obtained
with n-butylamine (n-BA) until the temperature was increased from 95°C to
170°C. The extent of delignification was about 50% at 170°C. The reducing sugar
obtained after enzymatic hydrolysis also increased by a factor of about 2. The
total sugar obtained by enzymatic hydrolysis was approximately the same
(30-32/100g) as that obtained after enzymatic hydrolysis of acid prehydrolyzed,
solvent-extracted hybrid poplar. If solvent extraction is to be used to pretreat
hybrid poplar, it appears that an initial H_2SO_4 prehydrolysis step is not necessary.

Sulphuric acid prehydrolyzed (95°C, 2% w/v H_2SO_4, 10% w/v slurry, 1h) corn
stover was also subjected to solvent extraction with various aliphatic and aromatic
alcohols (Lee et al, 1983). The extraction liquid consisted of 70% v/v solvent, 30%
v/v water with 0.5% w/v H_2SO_4 added as catalyst. The extraction temperature
was 160°C. For the aliphatic alcohols used (methanol, ethanol, propanol, butanol),
the extraction efficiency increased with increasing alcohol carbon number. For
example, 28% of prehydrolyzed corn stover was solubilized during extraction with
methanol, while 48% was solubilized during extraction with butanol. The aromatic
alcohols used (benzylalcohol, cyclohexanol, phenol), generally showed a more pronounced effect on the recovery of residual solid compared to the aliphatic alcohols. For most of the aromatic alcohols, about one-half of the prehydrolyzed corn
stover was solubilized. The results of 24h of enzymatic hydrolysis showed that the
reducing sugar (RS) produced was low (less than 12g RS/100g corn stover) for
most cases, as were the corresponding extents of conversion (less than 40%). In
fact, with the exception of n-butanol and phenol, the alcoholic extraction step produced results which were the same as, or poorer, in terms of fractional conversion,
than the control which was not subjected to solvent extraction.

It was observed from microscopic examination that the rigid cell structure of
corn stover was not appreciably disrupted even after the two-step acid and solvent pretreatments and enzymatic hydrolysis. In addition to the removal of lignin,
significant destruction of the rigid cellulose structure may be required to achieve a
high conversion of lignocellulosics to soluble sugars by processes incorporating
enzymic hydrolysis. Corn stover extracted with butanol (35% and 37%, respectively) and phenol (37%) showed a higher degree of conversion than was the case
with the other alcohols tested. Various amines, ethyleneglycol and its derivatives,

and dilute caustic solution were also examined for their effectiveness in the extraction of prehydrolyzed corn stover. n-Butylamine and ethanolamine extraction gave a relatively high degree of conversion (about 33%) upon enzymatic hydrolysis. Ethyleneglycol and its derivatives, methylcellosolve and ethylcellosolve, also were evaluated. Their effects on solid weight loss at 95 and $160\,^{\circ}C$ were found to be small. In all cases, a high degree of conversion during enzymatic hydrolysis was not achieved compared to the control.

Nghiem et al. (1984) reported on the delignification of corn stover with aqueous ethanol-NaOH mixtures. Corn stover was delignified with ethanol-water-NaOH mixtures and then hydrolyzed enzymatically. Runs were also carried out in which the corn stover was initially acid prehydrolyzed (2% w/v H_2SO_4, $95\,^{\circ}C$, 1h) prior to delignification. Table 2 shows the results of reducing sugar production from enzymatic hydrolysis of corn stover delignified with a 45% ethanol solution with and without prior acid pretreatment. At a given set of delignification conditions, the yield of sugars decreaseed with increasing caustic concentration for acid prehydrolyzed solids. Direct enzymatic hydrolysis of the delignified corn stover without prior acid prehydrolysis gave a higher sugar yield. The maximum yield of 77% based on the total carbohydrate content of the original sample was obtained at 75% delignification. The corresponding pretreatment conditions were 50% v/v ethanol, 4g NaOH/L, $95\,^{\circ}C$ for 4h, at which 90% of the carbohydrates remained insoluble.

Nghiem et al. (1984) found that the overall conversion yield of reducing sugar during enzymatic hydrolysis increased with increasing extent of delignification to a maximum yield of 77% at 75% delignification. Further increases in the extent of lignin removal resulted in reduced sugar yield presumably because of recrystallization of the cellulose. Nghiem et al. also found that enzymatic hydrolysis of delignified corn stover is more effective than enzymatic hydrolysis of pretreated, solvent delignified stover. This is in agreement with the findings of Lee et al. (1983) discussed earlier. The yield of reducing sugar obtained with the solvent pretreatment technique is too low to make the process economically attractive. In addition, solvent recovery costs may have a strong influence on the overall process economics. Little work has been done in this area. In general, solvent delignification pretreatment schemes are not as effective as gamma-irradiation or autohydrolysis low pressure steam pretreatment techniques.

4. Autohydrolysis Pretreatment

Lamptey et al. (1984a; 1985a) have presented details on the autohydrolysis steam pretreatment of corn stover and hybrid poplar without any explosion. Their results showed a significant effect of temperature on the yield of prehydrolyzed solids; the higher the temperature (up to $190\,^{\circ}C$), the more complete the removal of the hemicellulose fraction. The amount of by-products formed was found to increase with increasing temperature and processing time. The optimal conditions for the pretreatment of corn stover were found to be $190\,^{\circ}C$ and 30 min process time. Under these conditions, there was no apparent effect of solid-to-liquid ratio on the yield of prehydrolyzed solids. As expected, the reducing sugar content of the autohydrolyzed solids was found to increase with increasing slurry concentration. Mass balances for corn stover autohydrolysis showed that the hemicellulose and miscellaneous fractions were completely solubilized, while lignin and the cellulose fractions were practically undissolved.

The optimal conditions for the autohydrolysis of hybrid poplar were found to be $190\,^{\circ}C$ and 1h process time for slurry concentrations in the range 10-50% w/v. The yield of prehydrolyzed solids under these conditions was 67.5%. The longer process time required and the higher solids yield for autohydrolysis of hybrid pop-lar show the greater recalcitrance of wood to hydrolysis compared to corn stover. Comparison of autohydrolysis with H_2SO_4 pretreatment shows that autohydrolysis is a much more effective technique for the pretreatment of biomass. The results obtained by Lamptey et al. (1985) also show that autohydrolysis is a much more effective technique for the pretreatment of biomass than is gamma-irradiation. The pretreatment of corn stover using H_2SO_4 (Lee et al., 1983) under optimal con-ditions (2.0% w/v H_2SO_4, 95°C, 1h); gave a yield of prehydrolyzed solids of 60.3% compared to 53.2% for autohydrolysis. For hybrid poplar, Lee et al. (1983) obtained a solids yield of 72.5% after H_2SO_4 pretreatment (0.5% H_2SO_4, 1h). The amounts of by-products formed during the H_2SO_4 pretreatment were not determined. The difference between the effectiveness of autohydrolysis and the H_2SO_4 pretreat-ment is even more marked after enzymatic hydrolysis as discussed below.

The results of enzymatic hydrolysis of variously treated corn stovers over a 48-hour period presented by Lamptey et al. (1984a; 1985a) show that increasing the autohydrolysis temperature from 170°C to 190°C resulted in a significant increase in the rates and extents of saccharification for both simultaneous attri-tion-enzymatic hydrolysis and hydrolysis without attrition milling. Simultaneous attrition-enzymatic hydrolysis (hydrolysis with 60g glass beads/100 mL enzyme reaction mixture) of corn stover, previously subjected to autohydrolysis at 190°C for 30 min, yielded about 570 mg of soluble reducing sugar per gram of prehydro-lyzed solids within 24h. This represented about 95% saccharification of the input cellulose. At all the autohydrolysis temperatures investigated (170-190°C), signifi-cantly greater (about 40 to 50%) conversion to soluble sugars were obtained with attrition milling over hydrolysis runs without attrition milling. The extent of sac-charification obtained with the steam pretreatment (190°C, 30 min) was signifi-cantly greater (by about 30-35%) than the maximum value reported by Lee et al. (1983) for pretreatment with H_2SO_4.

For both corn stover and hybrid poplar, the results of enzymatic hydrolysis show that low-pressure steam autohydrolysis is more effective than the H_2SO_4 pretreatment technique in disrupting the structure of the lignocellulose complex with a resultant increase in the conversion of the cellulose to soluble sugars under optimal conditions, presumably because of the higher temperatures involved in autohydrolysis.

5. Enzymatic Hydrolysis of Pulpmill Sludge

Results obtained by Lee et al. (1983) and Lamptey et al. (1984b) showed that for an on-site bioconversion process unit, pulp mill sludge may be taken directly from a primary clarifier underflow stream and processed without drying in an agi-tated-bead enzymatic hydrolysis reactor up to concentrations of about 4% solids. The rates and extents of enzymatic hydrolysis of pulp mill sludge were found to be almost double for simultaneous attrition-saccharification over saccharification alone. About 64% conversion of the available total carbohydrate in pulp mill sludge was achieved within 24h.

6. Technoeconomics

Moo-Young et al. (1983, 1984, 1985b) and Lamptey et al. (1984c, 1985b) have presented detailed bioprocess technoeconomic comparisons on the production of fuel ethanol, methane biogas, and single cell protein from corn stover and pulpmill sludge. Based on current technology, sensitivity analyses of the three process options, carried out by Moo-Young et al. and Lamptey et al., revealed that ethanol production is uneconomical for North America, but methane fuel biogas production by anaerobic digestion and fungal SCP production by solid-substrate bioconversions indicate minimum economic plant sizes which seem to be practical in many North American locations.

CONCLUSIONS

Based on the Waterloo Biomass Group's results obtained to date, low-pressure steam autohydrolysis pretreatment of hybrid poplar and corn stover has proved to be a relatively simple and effective technique for the production of potential fermentable substrates. In terms of yield of prehydrolyzed solids, minimal by-product (sugar decomposition products) formation, and the rate and extent of the subsequent enzymic saccharification, the results of low-pressure steam autohydrolysis pretreatment were found to be as good as, or better than, those reported for more severe pretreatment processes. For hybrid poplar, the optimal conditions appear to be 190°C for 1h; and for corn stover, 190°C for 30 min. Under these conditions, a maximum sugar production of about 570 mg of soluble reducing sugar per gram of prehydrolyzed solids was obtained within 24h of enzymic saccharification. This value corresponds to about 95% saccharification of the input cellulose. The efficiency of the enzymatic hydrolysis of the autohydrolyzed solids was found to be dependent on the temperature and time used during the autohydrolysis pretreatment.

The next best pretreatment technique appears to be the combined sodium hydroxide-gamma irradiation pretreatment. The highest sugar yield of 94% (based on original total carbohydrates) after 48h enzymatic hydrolysis was obtained under the following pretreatment conditions: 100 Mrad and 0.06 g NaOH/g corn stover. However, it is to be noted that the method is highly susceptible to product (sugar) degradation.

From the organosolv experiments, aqueous ethanol-sodium hydroxide mixtures and n-butylamine were found to be most effective. A maximum sugar yield of 77% (based on original total carbohydrates) after 48h enzymatic hydrolysis was obtained at 75% delignification after pretreatment with 50% v/v ethanol, 4g NaOH/L, at 95°C for 4h. n-Butylamine, ethanolamine, ethylenediamine, ethyleneglycol, methylcellosolve and ethylcellosolve all gave sugar yields of less than 50-60% (based on original total carbohydrates) for both hybrid poplar and corn stover, although n-butylamine appeared to be the best. The relatively low yield of reducing sugar obtained with the organosolv technique may be attributed to both sugar degradation and the low degree of enzymatic conversion of solvent-extracted biomass. It is possible that extensive delignification results in recrystallization of the cellulose. With the aqueous ethanol-NaOH delignification studies, maximum sugar yield during enzymatic hydrolysis was obtained at about 75% delignification of the biomass. For both n-butylamine and aqueous ethanol-sodium hydroxide mixtures, it was found that direct enzymatic hydrolysis of the solvent-extracted biomass without prior acid prehydrolysis is much more effective in terms of sugar yield (higher by about 15-30%) than enzymatic hydrolysis of prehydrolyzed, solvent-extracted biomass.

From information reported in the literature, ball milling and the steam explosion process are also effective pretreatment techniques.

The criteria which appear to be crucial for the economic production of truly fermentable sugars from lignocellulosics using pretreatment, followed by enzyme and/or acid hydrolysis, are: (i) low raw material costs (function of plant siting); (ii) low pretreatment costs; (iii) quantitative conversion of all cellulosic materials to fermentable sugars, and (iv) utilization of all components of the residue, including the lignin.

Currently, there is virtually no pretreatment process that could be economically scaled-up to commercial scale. In most instances, the technical aspects of the production of fermentable sugars from lignocellulosics have been considered and the conceptual feasibility demonstrated. However, economic considerations will ultimately determine whether such schemes will actually be put into practice. Such factors as the cost of petroleum and the chemicals deriveable from it, the future conversion of coal into chemicals, and the cost of collecting, drying, trans- porting and storing lignocellulosic residues for enzymatic conversion will all have to be considered. Sensitivity economic analyses for ethanol, methane and SCP production from corn stover and pulpmill sludge, using current technology, indicate the unattractiveness of ethanol production at the present time.

ACKNOWLEDGEMENTS

This work was supported by a grant from the ENFOR Program, Canadian For- estry Service, Environment Canada. The authors wish to thank the Postdoctoral Fellows who worked on the project: Dr. J. Lamptey, Dr. M. Tanaka, Dr. N.P. Nghiem and Dr. Y.-H. Lee and Mr. Alejandro Gonzales-Valdes (Doctoral Student). The technical assistance of Mrs. Arlene Lamptey, Mr. Phil St. Amour and Mrs. Denise Salvian is gratefully acknowledged.

REFERENCES

Fan, L.T., Y.H. Lee and M.M. Gharpuray (1982). Adv. Biochem. Eng., 23, p. 157.

Gonzales-Valdes, A., M. Moo-Young and A. Lamptey (1981). Unpublished Results. University of Waterloo, Waterloo, Ontario, Canada

Humphrey, A.E. (1979) in: R.D. Brown and L. Jurasek (eds.). Hydrolysis of Cellulose: Mechanisms of Enzymatic and Acidic Catalysis, American Chemical Society, Wash- ington, D.C. p. 25.

Ladisch, M.R. (1979). Process Biochem., 21, January.

Lamptey, J. (1983). Doctoral Thesis. University of Waterloo, Waterloo, Ontario, Canada.

Lamptey, J., C.W. Robinson and M. Moo-Young (1984a). Proceedings 34th Canadian Chemical Engineering Conference. Canadian Society of Chemical Engineering, Quebec, Sept. 30.

Lamptey, J., M. Moo-Young and C.W. Robinson (1984b). Unpublished Results. Uni- versity of Waterloo, Waterloo, Ontario, Canada.

Lamptey, J., M. Moo-Young and P. Girard (1984c). Proc. of Int. Chemical Congress of Pacific Basin Societies. Hawaii, U.S.A. Dec. 16.

Lamptey, J., C.W. Robinson and M. Moo-Young (1985a). Enhanced enzymatic hydrolysis of lignocellulosic biomass pretreated by low-pressure steam auto- hydrolysis. Submitted to Biotechnol. Letts. (1985).

Lamptey, J., P. Girard and M. Moo-Young (1985b). Paper to be presented at 7th Int. Conf. on Global Impacts of Applied Microbiology (GIAM VII), Helsinki, Finland. Aug. 12-16.

Lee, Y.H., M. Moo-Young and C.W. Robinson (1983). Unpublished Results. University of Waterloo, Waterloo, Ontario, Canada.

Lee, Y.H., C.W. Robinson and M. Moo-Young (1983). Proceedings 33rd Canadian Chemical Engineering Conference, Canadian Society for Chemical Engineering, Toronto, Ontario Canada, Oct. 2-5.

Lee, Y.H., J. Nghiem, C.W. Robinson and M. Moo-Young (1984). Paper presented at the 7th Int. Biotechnol. Symp., New Delhi, India. February.

Moo-Young, M., J. Lamptey and P. Girard (1983). Proc. Symp. Biotechnology in the Pulp and Paper Industry, PIRA, London, U.K. Also in Press: Biotechnology Advances, Pergamon Press.

Moo-Young, M., J. Lamptey and P. Girard (1984). Proc. The First Arab Gulf Conference on Biotechnology and Applied Microbiology. Riyadh, Saudi Arabia, Nov. 12.

Moo-Young, M., C.W. Robinson and J. Lamptey. (1985a). ENFOR REPORT. DSS File No. 47SS-23216-3-6182, Contract Serial No. 0SS83-00030.

Moo-Young, M., J. Lamptey and P. Girard (1985b). Featured article. Chemical and Engineering News, p. 59. Jan. 14.

Muzzy, J.D., Roberts, R.S. and Fieber, C.A. (1982). Proceedings Feed, Fuels and Chemicals from Wood and Agricultural Residues Symp., ACS National Meeting. Sept. 14 Kansas City, MO.

Nghiem, N.P., A. Gonzales-Valdes, M. Moo-Young and C.W. Robinson (1984). Third European Congress on Biotechnology. Munchen, 10-14 September.

Phillips, C.R., Granatstein, D.G. and Wheatley, M.A. (1979). American Chemical Society Symposium Series, 90, 133-164.

A NEW APPROACH IN SOLID STATE FERMENTATION FOR CELLULASE PRODUCTION

D.S. Chahal
Bacteriology Research Centre
Institut Armand-Frappier
University of Quebec
531, boul. des Prairies
Laval, Quebec H7V 1B7

ABSTRACT

A cellulase system possessing high hydrolytic power was obtained by growing Trichoderma reesei by a new approach in solid state fermentation (SSF) of lignocelluloses. In this approach, the lignocelluloses are treated with sodium hydroxide to solubilize some of the hemicelluloses and lignin to expose cellulose. The solubilized hemicelluloses and lignin are retained in the fermentation medium. The lignocelluloses are also treated with a chemical-thermomechanical process for pulverizing into fibrous structure. Most of the hemicelluloses and lignin are also retained during this treatment. Cellulase yield of 200-430 IU/g cellulose was recorded by using this method. This is an increase of about 72% compared to yields of 100-250 IU/g cellulose in liquid state fermentation reported in the literature. The high cellulase activity 12-17 IU/mL and high cellulase yields (about 400 IU/g cellulose) were attributed to the growth of T. reesei on hemicelluloses during its early phase and enzyme production on cellulose during later phase as well as close contact of hyphae with substrate in SSF. This cellulase system was capable of hydrolyzing 78-90% of delignified wheat straw (10% concentration) in 96 hours without the addition of complementary enzymes, β-glucosidase and xylanases.

INTRODUCTION

The development of an economical process for ethanol (fuel) production through enzymatic hydrolysis of lignocelluloses is hindered because of the high costs of cellulase production, low cellulase activity per unit volume, and low concentration of sugar syrup obtained by hydrolysis of cellulose. A few processes for ethanol production have been reported in the literature, all of which suffer from some of these bottlenecks.

In the Berkeley process (23), only forty percent of the glucan is converted into a 2.6% sugar solution. The major problem in this process is that the concentration of sugar in the hydrolysate is very low (2.6%) for economical ethanol fermentation, therefore, it is concentrated from 2.6 to 11% with a multi-effect evaporator. In the Indian Institute of Technology (IIT) process (in ref. 3), about 71% of cellulose (Solka Floc) is converted into sugar syrup containing 20% cellobiose. The cellobiose is separated from glucose by fractional crystallization. The major drawback in this process is the presence of a large quantity of cellobiose in the hydrolysate. The use of a separate membrane cell and fractional crystallization to remove cellobiose from the hydrolysate would add extra cost to the process. In the Natick process (17) about 45% hydrolysable cellulose produces a 10% syrup of fermentable sugars. This process is more practical than others. The only drawback is the low conversion (45%) of cellulose to glucose.

A critical analysis of the literature on enzymatic hydrolysis has revealed that high cellulase activity per unit volume of fermentation broth is the most important factor in obtaining sugar concentrations of 20-30% from hydrolysis of cellulose for a process for ethanol production from cellulosic materials (3). It has also been confirmed that cellulase activity per unit volume can be increased by increasing the cellulose concentration in the medium (13). But it is not possible to handle more than 6% cellulose in a conventional fermenter because of rheological problems. In order to increase the cellulose concentration above 6%, solid state fermentation (SSF) seems to be the most attractive alternative (3,4). In the present report a new approach has been adopted to produce a complete cellulase system in SSF.

SOLID STATE FERMENTATION

Solid state fermentation (SSF) is defined as a process whereby an insoluble substrate is fermented with sufficient moisture, but without free water. In the liquid state or slurry state fermentation (LSF), on the other hand, the substrate is solubilized or suspended as fine particles in a large volume of water. Solid state fermentation is considered to require no complex controls and to have many advantages over the LSF (9), however, it has its own inherent problems (4). Of great historical relevance to modern technology is the "Koji" process i.e., growing of Aspergillus oryzae on rice (or other cereals) in the solid state. Production of "Koji" has been in practice in Japan since the eighth century for the production of Sake, the most traditional alcoholic drink in Japan.

Toyama (20) has done extensive work on cellulase production in SSF with various substrates. He held the opinion that wheat bran possesses all the factors for cellulase production on a commercial scale. The data given in Table 1 showed that when most of the starch and protein were removed from wheat bran, no decrease in cellulase activity was recorded. Moreover, when the wheat bran extract was added to pure cellulose (filter paper) very low cellulase activity was recorded. However, high cellulase activity was recorded on wheat-bran residue obtained after removal of starch and protein, and also some of the hemicelluloses and cellulose with cellulose enzyme, Meicelase. From this information he concluded that it was wheat bran residue which was responsible for cellulase production.

A new trend was observed when sawdust, rice straw and newspaper were delignified. The cellulase activity dropped considerably. This was not the case when corrugated cardboard was delignified (Table 2). Moreover, pure cellulose (filter paper) with wheat germ (containing starch and protein) gave very good cellulase activity, which was contrary to the results given in Table 1. Keeping in view all of the above noted observations, no concrete conclusion about the role of starch and protein, and delignification on cellulase production could be drawn.

THE NEW APPROACH

The new approach to produce cellulase complex with Trichoderma reesei is essentially a SSF similar to that of the Koji process of Toyama (20), except that wheat bran, wheat germ or rice bran, the expensive additives, are not used. Moreover, in the present approach, the lignocelluloses are not delignified since during delignification almost all of the hemicelluloses are removed. Instead, the lignocelluloses are treated with a small amount of

NaOH to solubilize some of the hemicelluloses and lignin to expose cellulose. The treated lignocelluloses are not washed and all the solubilized hemicelluloses and lignin are retained in the medium. Some NaOH-treatment is done with minimum quantity of water so that after the addition of nutrient solution and inoculum, the moisture content is less than 80% wt/wt without any free water in the medium (5). A new mutant QMY-1 of T. reesei was used in this study.

TABLE 1. Cellulase Production by
T. viride on Wheat Bran (20)

Organic Nitrogen Source		FPD-activity[a]
Wheat bran		666
Wheat bran sieved		666
Wheat bran extract	20 mL	333
plus filter paper	10 g	
Wheat bran residue		1,500

a   FPD-activity: Filter paper degrading activity (20).

TABLE 2. Effect of Delignification on Cellulase
Production by T. viride (20)

Solid Culture Medium	FPD-activity
Rice straw[a]	1,200
Delignified rice straw[b]	400
Newspaper	1,200
Delignified newspaper[c]	857
Corrugated cardboard	857
Delignified corrugated cardboard[c]	857
Filter paper	1,000

a   Petroyeast was added at 10% wt./wt. in rice straw and wheat germ was added at 20% wt./wt. in all the other substrates.
b   Delignified by boiling with a 1% NaOH solution for 3 h.
c   Delignified by boiling with a 20% peracetic acid solution for 1 h and then by autoclaving with a 1% NaOH solution at 120°C for 1 h.

i) Cellulase Production on Wheat Straw (WS)

Solid state fermentation was carried out with full concentration of nutrients in one set and with one-half concentration in the other set of experiments to evaluate the effect of different concentrations of salts in the medium, since some of the microorganisms are unable to grow in media with high osmotic pressure. The data presented in Table 3 indicated that T. reesei QMY-1 was quite tolerant to high concentration of nutrients. It produced the highest enzyme activity (8.6 IU/mL) and the highest cellulase yields (430 IU/g cellulose) on full concentration of nutrients in 22 days. However, when the nutrients were supplied in one-half concentration the cellulase activity dropped to 6.7 IU/mL and cellulase yield to 335 IU/g cellulose and there was no increase in enzyme yields after incubation of more than 14 days.

But, when the cultures were stirred after 11 days of growth and further incubated for 11 days without any stirring (total 22 days of incubation) the cellulase activity was reduced a little in the case of full concentration of nutrients whereas it increased considerably (8.0 IU/mL) in the case of one-half concentration of nutrients. It indicated that half of the quantity of required nutrients was sufficient to get optimum cellulase activity as well as cellulase yield. This would be a major contribution to reducing the cost of enzyme production.

When the concentration of WS was reduced to 5% (2% cellulose) by the addition of more water to make it LSF, the enzyme activity reached 6.0 IU/mL in 11 days with a yield of 300 IU/g cellulose. Which was equivalent to that of SSF at the same time of incubation. However, with further incubation in LSF the yield of cellulases decreased but in SSF the yield of cellulases continued to increase up to 22 days of inhibition.

The SSF proved to be better than LSF for cellulase production. The cellulase yields of 300 and 430 IU/g cellulose on WS in LSF and SSF, respectively, are quite high compared to that on pure cellulose in LSF reported for various mutants of T. reesei by other workers (Table 4). The increase in cellulase yields from a range of 160-250 IU/g of pure cellulose (Table 4) to the range of 300-430 IU/g crude cellulose (Table 3) from WS is considered due to the utilization of hemicelluloses during the initial growth of the organism and the production of cellulases on cellulose during the later phase of the growth.

ii) Cellulase Production on Aspen Wood Treated with NaOH

The data presented in Table 5 indicated that highest cellulase activity of 6.4 IU/mL and cellulase productivity of 194 IU/g cellulose were recorded after 20 days of growth of mutant QMY-1 on aspen wood (pretreated with 6% NaOH wt/wt and washed) in SSF. There was a slight decrease in cellulase production when the salt concentration was reduced to one-half. But cellulase production was reduced to a great extent (only 2.9 IU/mL) when the mutant QMY-1 was grown on alkali treated aspen wood in LSF. It again indicated that SSF was better than LSF for cellulase production. Low cellulase activity on alkali treated aspen wood in SSF and LSF as compared to that on alkali treated wheat straw might be due to removal of some of the hemicelluloses by washing after alkali treatment and difference in the physical and chemical nature of aspen wood as compared to that of WS.

However, cellulase activity (6.4 IU/mL) of QMY-1 on alkali treated aspen wood was quite higher than that of recent results on cellulase production by various mutants of T. reesei on Solka Floc (pure cellulose) and steam exploded wood (SE, SEWA) (Table 6).

TABLE 3. Cellulase Production on Wheat Straw[1] with Mutant QMY-1 (5)

Time of Incubation (Days)	Cellulase Activity (IU/mL)	Cellulase Yield (IU/g cellulose)
A. Solid State Fermentation (SSF) 2,3,4		
1. Full concentration of nutrients		
11	6.0	300
14	6.3	385
22	8.6	430
22[5]	7.4	370
2. One half concentration of nutrients		
11	5.5	275
14	6.7	335
22	6.7	335
22[5]	8.0	400
B. Liquid State Fermentation (LSF) 6		
Full concentration of nutrients		
7	1.3	65
11	6.0	300
14	5.5	275

1. Wheat straw was treated with 4% NaOH wt./wt. at 121°C for 0.5 h.
2. 5 g wheat straw + 20 g H_2O (no free H_2O) = 20% solids in each flask.
3. 5 g wheat straw contains 2 g cellulose ≅ 8% cellulose in each flask.
4. Cellulase enzyme was extracted by shaking the end product from each flask of SSF in 100 mL of water for 30 minutes on a shaker. Supernatant was obtained by centrifugation. The supernatant was used to test the cellulase activity. When the end-product of each flask of SSF was suspended in 100 mL H_2O for enzyme extraction, the concentration of substrate became 5% i.e. equivalent to LSF.
5. Cultures were stirred once after 11 days of incubation and were further incubated for 11 days without any stirring.
6. In LSF the pretreatment and cultural conditions were same as for SSF with full conc. of nutrients except that more sterile water was added to make 5% wt./vol. conc. of WS. The LSF cultures were incubated on shaker at 150 rpm, while the SSF cultures were kept still in humidified incubator at 30°C.

iii) **Cellulase Production on Aspen Wood Treated with Chemical-Thermomechanical Process (CTMP)**

The mild alkali treatment of CTMP did not make any difference for cellulase production by mutant QMY-1 (Table 7). Because the highest cellulase activity of about 6.3 IU/ml and cellulase productivity of 191 IU/g cellulose obtained on treated CTMP after 20 days in SSF was almost comparable to that obtained on untreated CTMP. The CTMP seems to be a good substrate for cellulase production even without any further treatment. Further studies on cellulase production on CTMP are in progress.

TABLE 4. Cellulase Production Potential of New Mutants of _Trichoderma_ _reesei_ (Liquid State Fermentation)

Mutant	Cellulose Concentration (%)	Cellulase Yields (IU/g cellulose)	Reference
QM 6a (Natick) (Parent strain)	6	100	(15)
QM 9414 (Natick)	0.75	240	(18)
	2	200	(8)
	2.5	172	(19)
	6	166	(15)
MCG 77 (Natick)	2	195	(8)
	6	183	(15)
NG 14 (Rutgers)	2	180	(8)
	6	250	(15)
C 30 (Rutgers)	5	290(?)	(19)
	5	160	(2)
	6	233	(15)
	15	200	(2)
L 27 (Cetus)	8	225	(16)
CL-847	5+1	229	(22)
D1-6	1	140	(14)

Range = 160 - 250

TABLE 5. Cellulase Production with Mutant QMY-1 in Solid State Fermentation and Liquid State Fermentation on Aspen Wood[1]

	Time (Days)	Cellulase Activity (IU/mL)	Cellulase Yields (IU/g Cellulose)
I.	**Solid State Fermentation (SSF)[2,3]** (20% wt./wt., no free water)		
(a)	Full concentration of salts		
	8	1.2	36
	12	2.1	64
	15	3.3	100
	20	6.4	194
(b)	One-half concentration of salts		
	8	0.6	18
	12	1.7	51
	15	3.2	97
	20	5.9	179
II.	**Liquid State Fermentation (LSF) (5% wt./wt.)**		
(a)	Full concentration of salts		
	8	0.2	6
	12	1.9	50
	15	2.5	76
(b)	One-half concentration of salts		
	8	0.4	12
	12	1.7	51
	15	2.9	88

1. Aspen wood was treated with 3 % NaOH solution in 1:2 (solid : liquid) ratio at 121°C for 1 h. Treated wood was washed to remove alkali which also removed some hemicelluloses and lignin.
2. 5 g Aspen wood (66% cellulose) in each flask.
3. Cellulase enzymes were extracted as explained for Table 1.

v) **Role of Hemicelluloses and Lignin on Cellulase Production**

To prove the above postulate, T. reesei QMY-1 was grown on pure cellulose in one set of experiments and cellulose was fortified with a mixture of hemicelluloses and lignin (solubles) obtained from WS in another set. The ratio of cellulose to WS solubles in the medium was 1:0.25 (6). Cellulase production was faster on pure cellulose as compared to that on fortified cellulose during the first phase of fermentation up to 85 h, thereafter, cellulase production increased considerably in the case of cellulose supplemented with WS solubles (Fig.1). Delayed and slow synthesis of cellulases during the early phase in the case of cellulose fortified with WS solubles was attributed to the presence of hemicelluloses, easily metabolizable carbon source. After the utilization of hemicelluloses, the cellulase synthesis increased considerably during the later phase of fermentation. This suggested that WS solubles which contained mostly hemicelluloses and lignin were responsible for high cellulase activity (3.4 IU/mL) and yields (340 IU/g cellulose).

Lignin and aromatic compounds related to lignin are reported to have an inhibitory effect on cellulases (1,21). The experiment was conducted with Solka Floc (SF) as cellulose source fortified with lignin (isolated with $NaClO_2$ from aspen wood) (6). The lignin was added at the rate of 40% of the dry weight of SF. This is the normal concentration of lignin in wood (i.e. 20% of wood and of 40 % cellulose). Contrary to the results obtained by other workers (1,21) no adverse effect of lignin (isolated with $NaClO_2$) was found on cellulase production. Cellulase activity (FPA) over 1.6 IU/mL was observed within 5 days (120 h) of fermentation of SF with or without lignin. Similarly, there was no adverse effect of lignin on β-glucosidase production (Fig.2). However, further work on the role of hemicelluloses and lignin individually and in combination is in progress.

TABLE 6. Cellulase Production by Different Mutants of T. reesei on Different Cellulosic Substrates (7)

Substrate (5%)	Cellulase Activity (IU/mL)	Cellulase Yields (IU/g Cellulose)
1. QM-9414		
Solka Floc	1.85	36.1
SE[2]	0.70	25.4
SEWA[3]	0.96	21.1
2. Rut C-30		
Solka Floc	5.56	111.0
SE	1.57	57.1
SEWA	2.10	46.1
3. E.58		
Solka Floc	1.93	38.6
SE	0.62	22.5
SEWA	0.51	11.2

1. Mutants were grown for 11 days at 28°C in LSF.
2. SE: Steam exploded wood (55% cellulose).
3. SEWA: Steam exploded wood water extracted and alkali treated (91%) cellulose).

v) **Composition of Cellulase System**

The cellulase system produced on WS in SSF contained the following enzymatic activities (IU/mL): cellulase 8.6, β-glucosidase 10.6, and xylanase 270 (5). The xylanase activity was quite variable (between 190 and 480 IU/mL); however, the ratio of cellulases and β-glucosidase varied between 1:1 and 1:1.5 in various cellulase system. These are enzyme activities when 5 g WS fermented in SSF was suspended in about 100 mL H_2O to extract the enzymes. High enzyme activities could be obtained by extracting the enzyme in a small quantity of water, i.e. double enzyme activity could be obtained by extracting enzyme in 50 mL H_2O. The composition of cellulase system indicated that there is no need for addition of extra β-glucosidase or xylanase for hydrolysis of pure cellulose or lignocelluloses. The addition of extra β-glucosidase in this cellulase system did not help to increase hydrolysis rate.

vi) Hydrolytic Potential

Wheat straw (WS) delignified by Toyama's method (20) was hydrolysed with cellulase system produced in SSF at 10% (wt /vol) consistency. Hydrolysis was done at pH 6.7, the

TABLE 7. Cellulase Production by Mutant QMY-1 on Aspen Pulp (CTMP) (5)

Substrate	Time (days)	Cellulase Activity (IU/mL)	Cellulase Yields (IU/g cellulose
(i) Treated CTMP	9	1.5	45
	16	4.7	142
	20	6.3	191
	26	6.0	182
	30	6.2	188
(ii) Untreated CTMP	9	2.2	67
	16	5.3	161
	20	5.0	151
	26	5.6	170
	30	7.2	218

1. Aspen wood was pulverized to fine fibers by a chemical-thermomechanical process (10). The pulp thus prepared is called chemithermomechanical pulp (CTMP). During this process most of hemicelluloses and lignin are retained.

2. Cellulose content of CTMP = 66 %

3. 5g dry wt. of substrate in each flask

4. Cellulase activity is measured as described under Table 1

5. Substrates were treated with 4 % NaOH (wt/wt) at 121°C for 1 h with 1:2 (solid:liquid) ratio. No washing.

66

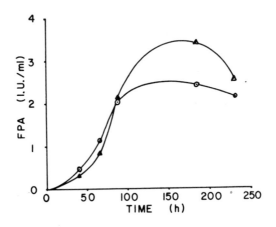

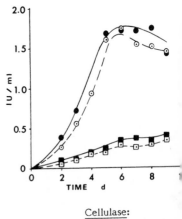

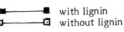

with hemicelluloses

without hemicelluloses

Fig. 1.   Effect of hemicelluloses
           on cellulase production (6)

Cellulase:

●————● with lignin
⊙————⊙ without lignin

β-glucosidase

■————■ with lignin
□————□ without lignin

Fig. 2.   Effect of lignin on
           cellulase and
           β-glucosidase
           production (6).

67

riginal pH of the enzyme solution (Fig. 3). The rate of hydrolysis of cellulose into glucose
as very high for the first 20 hours of hydrolysis, and about 65% of total hydrolysis was
ecorded. Almost all xylan and arabinan were hydrolysed within first the 20 hours of
ydrolysis. Very little cellobiose accumulated in the hydrolysate. Maximum concentration
f cellobiose accumulated was 7.75 g/L (Fig. 3) which is quite low to cause any inhibition on
ydrolysis of cellulose. After 20 hours of hydrolysis, the concentration of cellobiose
ecreased to about 3 g/L. The hydrolysate obtained after 96 h of hydrolysis contained
9.75 g sugars/L (glucose 68.18 g, cellobiose 3.19 g, xylose 26.71 g, and arabinose 1.67 g).

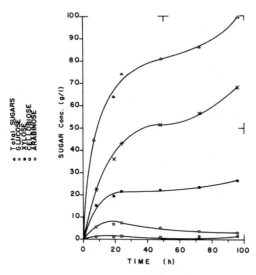

Fig. 3   Hydrolysis of delignified wheat straw.
40 IU of cellulase were used for each g
of delignified wheat straw at 45°C (5).

CONCLUSION

olid state fermentation seems to be most practical approach for the production of a
ellulase system containing all the complementary enzyme, β-glucosidase and xylanases,
ınd possessing high hydrolytic potential for hydrolysis of pure cellulose or lignocelluloses for
thanol (fuel) production. The high yields of the cellulase system and its high hydrolytic
otential have been attributed to the growth of T. reesei QMY-1 on hemicelluloses during its
arly phase and then on cellulose during its later phase for cellulase production as well as to
he close contact of hyphae with substrate in SSF. As the cellulase system is a complete
ystem for hydrolysis of lignocelluloses, therefore, there is no need for addition of extra
3-glucosidase and/or xylanases. There are also indications that the quantity of nutrients
:ould be reduced considerably. Reducing nutrient quantity in the fermentation medium and
10 addition of β-glucosidase and/or xylanases will reduce the costs of enzymatic hydrolysis
:onsiderably. However, SSF could have its own inherent problems, i.e. maintenance of pH,
noisture level, aeration, etc., when handling large quantities of solid substrates (4).

REFERENCES

1. Arrieta-Escobar, A. and Belin, J.M. 1982. Effects of phenolic compounds on the growth and cellulolytic activity of a strain of Trichoderma viride. Biotechnol. Bioeng. 24: 983-989.

2. Blanch, H.W. and Wilke, C.R. 1982. Cellulase production and kinetics. Proc. Internl. Symp. on Ethanol from Biomass. Winnipeg, Canada. October 13-15 pp. 415-441.

3. Chahal, D.S. 1982. "Enzymatic Hydrolysis of Cellulose: State-of-the-Art". Division of Energy Research and Development, National Research Council Canada, Ottawa, Canada, K1A 0R6, Report, pp. 74.

4. Chahal, D.S. 1982. Growth characteristics of microorganisms in solid state fermentation for upgrading protein values of lignocelluloses and cellulase production. In: Foundation of Biochemical Engineering: Kinetics and Thermodynamics in Biological Systems. ACS Symp. No. 207, 421-442.

5. Chahal, D.S. 1985. Solid-State fermentation with Trichoderma reesei for cellulase production. Appl. Environ. Microbiol. 49, 205-210.

6. Chahal, D.S. 1984. Process Development for Enzymatic Hydrolysis of Cellulose. Division of Energy Research and Development, National Research Council Canada, Ottawa, Canada, K1A 0R6, Final Report.

7. ENFOR Project C-181 (2). 1983. Conversion of Cellulose to Ethanol Using a Two-stage Process. Forintek Canada Corp. Ottawa, Ontario.

8. Gallo, B.J., Andreotti, R., Roche, C., Ruy, D., and Mandels M. 1978. Cellulase production by a new mutant strain of Trichoderma reesei MCG77. Biotechnol. Bioeng. Symp. No. 8, 89-101.

9. Hesseltine, C.W. 1976. Solid State Fermentations. Biotechnol. Bioeng. 14, 517-532.

10. Law, K.N., Lapointe, M. and Valada, J.L. 1983. Production of CTMP from aspen. International Mechanical Pulping Proc. (Tappi Proc.) 259-270.

11. Mandels, M., Medeiros, J.E., Andreotti, R.E. and Bisset, F.H. 1981. Evaluation of cellulase culture filtrates under use conditions. Biotechnol. Bioeng. 23, 2009-2026.

12. Mandels, M. and Weber, J. 1969. The production of cellulases. Adv. Chem. Ser. 95, 391-414.

13. Nystrom, J.M. and DiLuca, P.H. 1977. Enhanced production of Trichoderma cellulase on high levels of cellulose in submerged culture. Proc. of Bioconversion Symp. I.I.T., New Delhi, 293-304.

14. Panda, T., Bisaria, V.S. and Ghose, T.K. 1983. Studies on mixed fungal culture for cellulase and hemicellulase production part-I: Optimization of medium for the mixed culture of Trichoderma reesei D1-6 and Aspergillus wantii Pt. 2804. Biotechnol. Letters 5, 767-772.

15. Ryu, D.D. and Mandels, M. 1980. Cellulases: biosynthesis and applications. Enzyme Microb. Technol. 2, 91-101.

69

16. Shoemaker, S.P., Raymond, J.C. and Bruner, R. 1981. Cellulases: Diversity among improved Trichoderma strains. In: Trends in the Biology of Fermentations for Fuels and Chemical. Eds. A. Hollaender et al. Plenum Publishing Corp. New York, pp. 89-109.

17. Spano, L., Allen, A., Tassinari, T., Mandels, M. and Ryu, D. 1978. Proc. 2nd Annu. Symp. Fuels from Biomass. Ed. W.W. Shuster, John Wiley and Sons, New York, N.Y.

18. Sternberg, D. 1976. Production of cellulase by Trichoderma. Biotechnol. Bioeng. Symp. No. 6, 35-53.

19. Tangnu, S.K., Blanch, H.W., and Wilke, C.R. 1981. Enhanced production of cellulase, hemicellulase and β-glucosidase by Trichoderma reesei (Rut C30). Biotechnol. Bioeng. 23, 1837-1849.

20. Toyama, N. 1976. Feasibility of sugar production from agricultural and urban cellulosic wastes with Trichoderma viride cellulase. Biotechnol. Bioeng. Symp. No. 6, 207-219.

21. Vohra, R.M., Shirkot, C.K., Dhawan, S. and Gupta, K.G. 1980. Effect of lignin and some of its components on the production and activity of cellulase(s) by Trichoderma reesei. Biotechnol. Bioeng. 22: 1497-1500.

22. Warzywoda, M., Ferre, V. and Pourquie, J. 1983. Development of a culture medium for large scale production of cellulolytic enzyme by Trichoderma reesei. Biotechnol. Bioeng. 25: 3005-3010.

23. Wilke, C.R., Yang, R.D., Sciamanna, A.F. and Freitas, R.P. 1978. Raw materials evaluation and process development studies for conversion of biomass to sugars and ethanol. Proc. 2nd Annu. Symp. on Fuels from Biomass, Dept. of Energy Meeting, Troy, N.Y. June 20-22.

PRODUCTION AND USE OF CELLULASES IN THE CONVERSION OF CELLULOSE TO FUELS AND CHEMICALS

A.W. Khan

Division of Biological Sciences
National Research Council of Canada
Ottawa, Ontario, Canada   K1A 0R6

ABSTRACT

To improve the enzymatic hydrolysis of cellulosic biomass to fermentable sugars, the cellulase production is being studied in three different systems. These systems are (a) the simultaneous production of cellulases, saccharification and fermentation for the conversion of cellulose to fuels and chemicals; in a single step by the use of cocultures, (b) simultaneous cellulase production and saccharification by the use of mesophilic anaerobes and (c) cellulas production by Trichoderma reesei for subsequent hydrolysis of cellulose by these enzymes. By simultaneous cellulase production ar fermentation, it appears possible to hydrolyze both cellulose and hemicellulose, and convert all the three sugars produced namely cellobiose, glucose and xylose to ethanol in a single step. Results of recent work show that, in cellulolytic anaerobes, cellulolysis occurs as a result of cell-associated cellulases rather than extracellular cellulases, and cell-associated cellulases are responsible for the accumulation of sugars in culture media. The work carried out on cellulase production by T. reesei RUT. C-30 indicates that this strain is capable of producing enzyme concentrations upto 30 IU/ml of filter paper activity.

INTRODUCTION

Recent work carried out in our laboratory has led to the development of a number of fermentation processes for the conversion of cellulosic biomass to chemicals and fuels, and has demonstrated the usefulness of these processes in principle (Khan et al. 1981, 1983 and 1984a). These processes are based on simultaneous production of cellulolytic enzymes, saccharification and fermentation. Since the efficiency of these processes depends on t conversion of cellulose to sugars, the study of cellulolytic enzyme production forms an important part of this work. This study is bei carried out with both cellulolytic anaerobes and cellulase producin fungi, for simultaneous production of enzymes and saccharification, as well as for sequential enzyme production and saccharification. This paper outlines (a) the development of a fermentation process f simultaneous production of cellulases, saccharification of both cellulose and hemicellulose, and conversion of cellobiose, glucose and xylose to ethanol, (b) production of cellulases by cellulolytic anaerobes and (c) production of cellulases by Trichoderma reesei.

ULTS AND DISCUSSION

) Simultaneous production of cellulases, saccharification and
conversion of sugars to ethanol:

For the direct conversion of cellulosic biomass to ethanol,
coculture consisting of two mesophilic anaerobes, one cellulolytic
1 the other ethanologenic was developed (Khan and Murray, 1982).
is coculture produced about 0.8-1.0 mole of ethanol per mole of
llulose (as glucose equivalents) from a variety of cellulosic
terials. The cellulolytic anaerobes used in this coculture have
en isolated in our laboratory, and are, Acetivibrio cellulolyticus
atel et al. 1980), Acetivibrio cellulosolvens (Khan et al. 1984)
1 Bacteriodes cellulosolvens (Murray et al. 1984). These anaerobes
oduce cellulolytic enzymes capable of degrading both cellulose and
micellulose. The cellulose is degraded into cellobiose and
ucose, and hemicellulose to xylose. The cellobiose is produced in
antities higher than these anaerobes can utilize, while glucose and
lose are not utilized as a carbon source by these anaerobes, and
nsequently these sugars accumulate in the broth. The ethanologenic
aerobe, Clostridium saccharolyticum used in this coculture has the
ility to convert all these sugars to ethanol. The strain
provement work on C. saccharolyticum has led to the development of
tants capable of producing higher concentrations of ethanol and
lerating ethanol levels as high as 7-10% (Asther and Khan, 1984b,
rray et al. 1983). Although C. saccharolyticum has the ability to
nvert cellobiose, glucose and xylose to ethanol, it perfers glucose
d in mixtures containing these sugars, the rate of conversion of
llobiose and xylose to ethanol is slow. To overcome this
fficulty, we have successfully developed a coculture consisting of
momonas anaerobia and C. saccharolyticum (Asther and Khan, 1984a).
this coculture, Z. anaerobia was used to remove glucose by
nverting it to ethanol. The remaining two sugars, namely
llobiose and xylose are converted to ethanol by C. saccharolyticum.
ing this coculture, over 4% ethanol was produced in 4 days from a
nthetic mixture containing 26 g/L of cellobiose, 78 g/L of glucose
d 26 g/L of xylose (Table 1). A similar coculture has been
ccessfully used for the production of ethanol from sugars obtained
hydrogen-fluoride treatment of aspen wood (Murray and Asther,
84). The possibility of using this coculture together with
llulolytic anaerobes for simultaneous cellulase production;
ccharification and ethanol fermentation of cellulose; as well as
r producing ethanol from sugars obtained by the action of
llulases of Trichoderma has also been demonstrated. For sequential
ccharification and fermentation, a relation between filter paper
tivity and saccharifying ability of the Trichoderma enzyme system
orisset and Khan, 1984) has been established and over 6% sugar
lutions have been obtained (Murray and Duff, 1984).

) Cellulase production by mesophilic anaerobes:

Mesophilic anaerobes recently isolated in our laboratory
ve been shown to produce both extracellular and cell-associated
llulases (Saddler and Khan, 1980) and accumulate fermentable sugars
the broth (Giuliano et al. 1983). However, the extracellular
zymes produced by these anaerobes are less than 1/10th the amount

Table 1. Ethanol production by the coculture from a mixture containing 20% cellobiose, 60% glucose and 20% xylose under pH controlled conditions[a]

Microorganism	Incubation time (days)	Sugars used (g.L^{-1})	(% of the initial conc.)	Ethanol produced (g.L^{-1})	(mole/mole monosacchar used)
Z. anaerobia and	2	82.0	63.1	38.6	1.8
C. saccharolyticum,	4	93.2	71.7	42.8	1.8
A-15	7	126.7	97.5	51.4	1.6

[a]Medium contained 130 g.L^{-1} of total sugars.

Asther and Khan 1984b.

Table 2. Enzymatic saccharification of various cellulosic materials and their conversion to ethanol by a coculture of C. saccharolyticum and Z. anaerobia[a].

Cellulosic material[b]	Total reducing sugars (g/L)	Saccharification[c] (%)	Ethanol produced (g/L)	Ethanol produc (mole/mole o sugar)
Solka-floc	35.5	71	15.4	1.8
Cellulose CF11	28.8	58	13.1	1.8
SED[e]	16.2	45	7.0	1.7
SED washed[e,f]	23.2	52	10.4	1.8

[a]Incubation time, 48 hr at 37°C.

[b]Initial concentration, 50 g/L.

[c]Based on cellulose content of the material, saccharification time, 48 hr at 50°

[d]Expressed in glucose equivalent.

[e]Steam and explosion decompressed aspen wood.

[f]Washed with 0.4% NaOH solution.

(Khan and Asther 1984. Poster, 7th Int. Biotech. Symp, New Delhi, India, Feb. 19-25.

roduced by the Trichoderma and do not account for the cellulolytic
ctivity of these anaerobes or their ability to accumulate sugars in
arge amounts. More recent work carried out in our laboratory has
hown that the cellulolytic enzymes associated with cells are mainly
esponsible for sugar formation (Giuliano and Khan, 1984), and that
heir separation from the cells by conventional methods lead to the
oss of activity. We have obtained sugar concentrations as high as
5 g/L by incubating washed cells with cellulose at pH 5.0. These
indings indicate that cellulolytic enzyme system produced by these
naerobes is different than that produced by fungi, and newer
pproaches are being tested to utilize these cellulolytic anaerobes
or saccharification.

Table 3. Highest cellulase yields obtained in batch culture and in fed-batch
fermentation by two highly productive strains of Trichoderma reesei.

Strain	Substrate used (%)	Enzyme produced (IU FP/ml)	Protein produced (mg/ml)	Source of information
Batch culture				
C-30	Cellulose (3-4)	4-6	6-7	Routinely obtained
C-30	Cellulose (5-10)	8	-	Hendy et al. (1982)
C-30	Cellulose (5)	9	14	Warzywoda et al. (1983)
CL-847	Cellulose (5) and wheat bran (2)	17	22	Warzywoda et al. (1983)
Fed-batch culture				
C-30	Filter paper	30	47	Hendy et al. (1982, 1984)
CL-847	Lactose and pulp	23	37	Pourquie et al. (1984)
CL-847	Lactose	25-30	35-40	Duran et al. (1984)

c) Production of cellulolytic enzymes by Trichoderma reesei:

It is only recently that the work on cellulase production
y T. reesei was started in our laboratory. This work is being
arried out using T. reesei RUT. C-30 strain. This strain was
elected for its high productivity amongst other hyper-cellulase
roducing strains available for research. The highest yields of
ilter paper (FP) activity obtained in different laboratories from
wo hyper-cellulase producing strains of T. reesei are shown in Table
. In batch culture, enzyme concentrations upto 8-9 IU/ml of FP

74

activity with strain C-30 and 17 IU/ml with strain CL-847 have been
reported (Hendy et al. 1982, Warzywoda et al. 1983). In shake
flasks, we are routinely getting about 4-6 IU/ml of FP activity. The
work in progress in our laboratory suggests that enzyme production
can be enhanced considerably by the addition of chemicals capable of
reducing catabolite repression. Using a fed-batch fermentation
technique FP activities as high as 23-30 IU/ml have been reported
(Hendy et al. 1982 and 1984, Pourquie and Vandecasteele, 1984). The
fermentation time required to obtain these yields is about 5-7 days
for strain CL-847 (Duran et al. 1984) and about 6-10 days for strain
C-30 (Hendy et al. 1982 and 1984). The enzyme system produced by
CL-847 is also richer in β-glucosidase and xylanase activities
(Warzywoda et al. 1983). However it has been pointed out by
Warzywoda et al. (1983) that the media being used to obtain these
high yields cannot be scaled up due to their cost and high viscosity,
and evaluation of these yields on realistic media is necessary.

REFERENCES

Asther, M. and Khan, A.W. 1984a. Influence of the presence of
    Zymomonas anaerobia on the conversion of cellobiose, glucose and
    xylose to ethanol by Clostridium saccharolyticum. Biotech.
    Bioeng. 16: 970-972.

Asther, M. and Khan, A.W. 1984b. Improved fermentation of cellobiose,
    glucose and xylose to ethanol by Zymomonas anaerobia and a high
    ethanol tolerant strain of Clostridium saccharolyticum. Appl.
    Microbiol. Biotechnol. 20: in press.

Durand, H., Clanet, M. and Tiraby, G. 1984. A genetic approach of the
    improvement of cellulase production by Trichoderma reesei.
    Poster presented at Bioenergy 84. Goteborg, Sweden, June
    18-21.

Giuliano, C., Asther, M. and Khan, A.W. 1983. Comparative degradation
    of cellulose, and sugar formation by three newly isolated
    mesophilic anaerobes and Clostridium thermocellum. Biotechnol.
    Lett. 5: 395-398.

Giuliano, C. and Khan, A.W. 1984. Cellulase and sugar formation by
    Bacteriodes cellulosolvens, a newly isolated cellulolytic
    anaerobe. Appl. Environ. Microbiol. 48: 446-448.

Hendy, N.A., Wilke, C.R. and Blanch, H.W. 1984. Enhanced cellulase
    production in fed batch culture of Trichoderma reesei C30.
    Enzyme Microb. Technol. 6: 73-77.

Hendy, N., Wilke, C. and Blanch, H. 1982. Enhanced cellulase
    production using solka floc in fed batch fermentation.
    Biotechnol. Lett. 4: 785-788.

Khan, A.W., Asther, M. and Giuliano, C. 1984a. Utilization of steam
    and explosion decompressed aspen wood by some anaerobes. J.
    Ferment. Technol. 62: in press.

Khan, A.W., Meek, E., Sowden, L.C. and Colvin, J.R. 1984b. Emendation of the Genus Acetivibrio and description of Acetivibrio cellulosolvens sp. nov., a non-motile cellulolytic mesophile. Int. J. Syst. Bacteriol. 34: in press.

Khan, A.W., Miller, S.S. and Murray, W.D. 1983. Development of a two-phase combination fermenter for the conversion of cellulose to methane. Biotech. Bioeng. 15: 1571-1579.

Khan, A.W., Wall, D. and van den Berg, L. 1981. Fermentative conversion of cellulose to acetic acid and cellulolytic enzyme production by a bacterial mixed culture. Appl. Environ. Microbiol. 41: 1214-1218.

Morisset, W.M.L. and Khan, A.W. 1984. Relation between filter paper activity and saccharifying ability of a Trichoderma cellulase system. Biotechnol. Lett. 6: 375-378.

Murray, W.D. and Asther, M. 1984. Ethanol fermentation of hexose and pentose wood sugars produced by hydrogen-fluoride solvolysis of aspen wood. Biotechnol. Lett. 6: 323-326.

Murray, W.D. and Duff, S.J.B. 1984. Studies on the production and use of cellulases for the conversion of cellulose to fermentable sugars and alcohol. Proc. 6th Int. Symp. on Alcohol Technology, pp. 179-183. Ottawa, May 21-25.

Murray, W.D., Sowden, L.C. and Colvin, J.R. 1984. Bacteriodes cellulosolvens sp. nov., a cellulolytic species from sewage sludge. Int. J. Syst. Bacteriol. 34: 185-187.

Murray, W.D., Wemyss, K.B. and Khan, A.W. 1983. Increased ethanol production and tolerance by a pyruvate-negative mutant of Clostridium saccharolyticum. Eur. J. Appl. Microbiol. Biotechnol. 18: 71-74.

Patel, G.B., Khan, A.W., Agnew, B.J. and Colvin, J.R. 1980. Isolation and characterization of an anaerobic cellulolytic microorganism, Acetivibrio cellulolyticus, gen. nov., sp. nov. Int. J. Syst. Bacteriol. 30: 179-185.

Pourquie, J. and Vandecasteele, J.P. 1984. Substituted fuels production from lignocellulosics: Cellulolytic enzyme production. Poster presented at Bioenergy 84. Goteborg, Sweden, June 18-21. Abstract pp. 176.

Saddler, J.N. and Khan, A.W. 1980. Cellulase production by Acetivibrio cellulolyticus. Can. J. Microbiol. 26: 760-765.

Warzywoda, M., Vandecasteele, J.P. and Pourquie, J. 1983. A comparison of genetically improved strains of the cellulolytic fungus Trichoderma reesei. Biotechnol Lett. 5: 243-245.

Bacterial Cellulases

C.R. MacKenzie
Division of Biological Sciences
National Research Council of Canada,
Ottawa, Ontario, Canada K1A OR6

Certain prokaryotes, especially some anaerobes, are capable of rapidly degrading crystalline cellulose in vivo. However, attempts at duplicating these rates in vitro have been largely unsuccessful. The cellulase system of Acetivibric cellulolyticus has been investigated in an effort to understand these discrepancies It has been observed that removal of spent culture medium is a prerequisite for efficient hydrolysis of native cellulose in vitro. In addition a requirement for Ca^{++} has been demonstrated. It was further shown that Ca^{++} exercised this effect b stabilizing cellulase activity. Also, thiol-binding reagents inhibited enzyme activity suggesting the involvement of thiol-groups in cellulose solubilization. A second organism, Streptomyces flavogriseus, has been studied as a representative of the actinomycetes - another group of prokaryotes capable of efficient growth on cellulose. The cellulase complex produced by this organism was found to more closely resemble fungal systems such as that of Trichoderma reesei than those of other bacterial cellulose degraders. It has low specific activity, contains severa endoglucanases and an exoglucanase accounting for a large part of the total protein in cellulase preparations.

INTRODUCTION

Compared to cellulolytic fungi such as Trichoderma reesei, bacterial cellulase producers remain relatively unstudied. In recent years however, there ha been increased interest in these organisms - particularly the cellulolytic anaerobes. Many studies have pointed out the highly efficient metabolism of cellulose by these organisms. However, doubts have been expressed regarding the usefulness of these organisms in saccharification processes because of failure to translate the high in vivo hydrolysis rates into in vitro situations. In attempts at explaining this discrepancy, conjecture has centered around the possible disruption of a highly ordered cellulase complex which may exist in growing cultures. While this may be partially true, there are other factors which play an important role - at least in A. cellulolyticus, a bacterial anaerobe.

EXPERIMENTAL

The rates of cellulose digestion and end-product formation by Acetivibrio cellulolyticus during optimized growth on Avicel, a microcrystalline form of cellulose have been reported previously (Patel and MacKenzie, 1982). The maximum substrate utilization rate (0.15g/l/h) compares very favourably with other cellulolytic organisms - both bacteria and fungi. Endoglucanase levels of approximately 0.4U/ml were detected in the supernates of such cultures. In contrast, endoglucanase levels of 150U/ml have been reported for hyperproducing strains of Trichoderma reesei. The results obtained for Avicel hydrolysis by culture supernates also showed enormous activity differences between these two organisms. The supernates from A. cellulolyticus cultures grown on Avicel showed negligible activity towards their substrate.

Protein was precipitated from A. cellulolyticus culture supernates with ammonium sulphate and desalted by Bio Gel P-6 (Bio-Rad laboratories) chromatography. Hydrolysis of Avicel and acid-swollen cellulose by enzyme preparations was followed by the decrease in turbidity (measured at 660 nm) in a manner similar to that described by Johnson et al. (1982). Unlike culture supernates, desalted preparations demonstrated an ability to hydrolyze Avicel to a significant extent Fig. 1) . Avicel hydrolysis rates by desalted A. cellulolyticus enzyme and T. eesei were compared at activity levels (based on endoglucanase) similar to those ncountered in culture supernates. In spite of the vast differences in protein and nzyme concentration, the hydrolysis rates were practically identical.

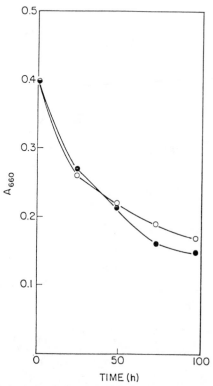

TIME (h)

ig 1. Hydrolysis of 0.1% Avicel by Acetivibrio cellulolyticus (•) and T. reesei (o) cellulase at endoglucanase levels of 0.4U/ml and 150U/ml respectively. Hydrolysis mixtures also contained 50 mM phosphate buffer, pH 6.

Two metal ions, Ca^{++} and Hg^{++}, were observed to have a profound effect on vicel hydrolysis by A. cellulolyticus cellulase (Fig. 2). The enzyme was ompletely inactive towards Avicel in the absence of Ca^{++} with concentrations as low s 1 mM giving linear hydrolysis rates following a more rapid initial drop in urbidity. Increasing the Ca^{++} concentration up to 10 mM did not give further ncreases in hydrolysis rate. Because of the absolute requirement for Ca^{++}, the

effect of Hg^{++} could only be tested in the presence of both ions. The addition o
Hg^{++} at a concentration of 10 µM resulted in a hydrolysis rate which was
approximately 50% of that observed in control assays containing Ca^{++}. Addition o
Hg^{++} at 24h following the initial rapid hydrolysis rate almost completely inhibit
further hydrolysis. The inhibition of cellulolysis by Hg^{++} suggests the involvem
of sulfhydryl groups in the hydrolysis process. These results confirm earlier wo
on A. cellulolyticus cellulase (MacKenzie and Bilous, 1982). A requirement for
reduced sulfhydryl groups has also been demonstrated in Clostridium thermocellum,
another cellulolytic anaerobe (Johnson and Demain, 1984). The effects of these ic
on acid-swollen cellulose hydrolysis were somewhat different. While Hg^{++} reduced
the hydrolysis rate, Ca^{++} was not observed to significantly increase the rate.

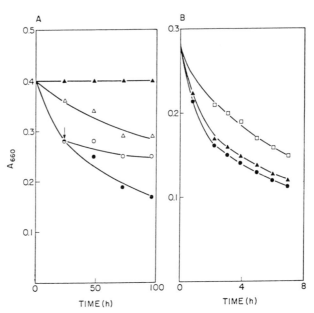

Fig. 2. Effect of Ca^{++} and Hg^{++} on the hydrolysis of 0.1% Avicel (A) and 0.25%
acid-swollen cellulose (B) by Acetivibrio cellulolyticus cellulase. No
additions (▲); 10 mM Ca^{++}(o); 10 µM Hg^{++} (□); 10 mM Ca^{++} and 10 µM Hg^{++}(∆
The inhibitory effect of 10 µM Hg^{++} on Avicel hydrolysis was also tested
its addition at 24h (↓) to a hydrolysis mixture containing 10 mM Ca^{++}.
Hydrolysis mixtures also contained 50 mM phosphate buffer, pH6, and 0.4
units/ml of endoglucanase.

The susceptibility of the A. cellulolyticus cellulase complex to
inhibition by glucose and cellobiose is shown in Figure 3. Distinctly different
inhibition profiles were obtained for Avicel and acid-swollen cellulose hydrolysis
This suggests that different enzymes are involved in the hydrolysis of these two
substrates. With each substrate cellobiose was much more inhibitory than glucose.
Although A. cellulolyticus produces small amounts of cellulase compared
fungal cellulose degraders, the number of electrophoretically distinct forms is
large. A Congo Red activity stain, performed as described previously (MacKenzie a

79

lliams, 1984), revealed the presence of at least seven distinct endoglucanases
ig. 4). Three endoxylanases could also be detected by this technique. In
oducing such a complex array of endoglucanase activities, A. cellulolyticus is
milar to Clostridium thermocellum, another bacterial anaerobe. In
thermocellum, theseactivities apparently exist in the form of high molecular
ght aggregates which bind to cellulose by means of cellulose-binding factors.
cellulose-binding factor has been identified as a protein with a molecular
ght of 210,000 daltons; it is apparently devoid of endoglucanase activity (Lamed
al., 1983).

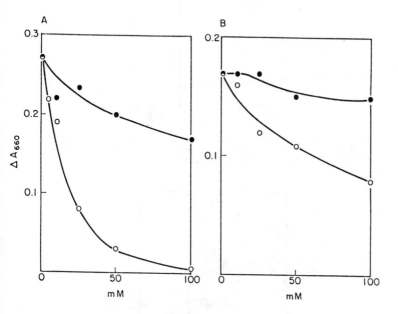

3. Effect of glucose (●) and cellobiose (o) on the hydrolysis of 0.1% Avicel
(A) and 0.25% acid-swollen cellulose (B) by Acetivibrio cellulolyticus
cellulase. Hydrolysis mixtures contained 50 mM phosphate buffer, pH6, and
0.4 units/ml of endoglucanase.

Although the enzyme preparations used in the hydrolysis and
ectrofocusing studies were derived from culture supernates, there is some evidence
ch indicates that some of the cellulase activity is localized on the cell
face. We have previously reported that approximately 10% of the cellulase
ivity (measured with cellulose-azure as the substrate) was cell-associated
cKenzie and Bilous, 1982). The data presented in Table 1 indicate that 50% of
cells present in cultures actively growing on cellobiose readily adsorbed to
lulose as evidenced by their removal from cell suspensions following the addition
Avicel PH105. This is slightly lower than the value of 70% adsorption to
lulose which has been

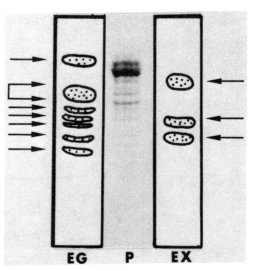

Fig. 4. Protein (P), endoglucanase (EG) and endoxylanase (EX) stains of isoelectri
focused <u>Acetivibrio cellulolyticus</u> enzyme. Samples were electofocused and
stained <u>as described</u> previously (MacKenzie and Williams, 1984).

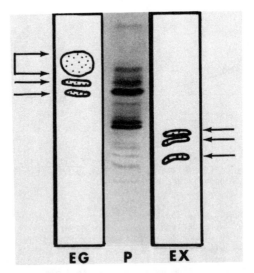

Fig. 5. Protein (P), endoglucanase (EG) and endoxylanase (EX) stains of isoelectri
focused <u>Streptomyces flavogriseus</u> enzyme. Samples were electrofocused as
described <u>previously</u> (MacKenzie and Williams, 1984).

Table 1
Adsorption of Cells to Cellulose

Treatment	A. cellulolyticus		Clostridium thermocellum[‡]	
	ΔA_{660}	% Reduction	ΔA_{660}	% Reduction
Control 30 min without cellulose	0.05	6	--	--
1% Avicel PH105 (30 min)	0.43	51	0.49	70

‡ Data from Bayer et al., J. Bacteriol. 156: 818, 1983.

reported for C. thermocellum. In C. thermocellum, this adsorption has been
attributed to cellulose-binding factors located on the cell surface (Bayer et al.,
1983; Lamed et al., 1983). Studies on cellulolysis by Streptomyces flavogriseus
45-CD, an actinomycete, revealed the existence of a cellulase complex which more
closely resembled fungal systems such as that of Trichoderma reesei than those of
bacterial anaerobes (MacKenzie et al., 1984, 1984a). In contrast to the anaerobes
large amounts of cellulase protein were produced. Fractionation of the complex
yielded six endoglucanase fractions and a major exoglucanase fraction accounting for
a large percentage of the total protein (MacKenzie et al., 1984). Protein and
activity (endoglucanase and endoxylanase) stains of isoelectric focused
S. flavogriseus enzyme are shown in Figure 5. All of the endoglucanases focused in
the pH 3.5-4.5 range and were not completely resolved in the activity stain. This
technique emphasized the enormous difference in activity levels found in culture
supernates of this organism and those of Acetivibrio cellulolyticus. While culture
supernates from the latter organism were in an undiluted form those of S.
flavogriseus were diluted 1:100. The endoxylanase stain showed the existence of
three activities in the S. flavogriseus enzyme. These enzymes were distinct from
the endoglucanases.

CONCLUSIONS

1. There is a good possibility that further research on efficient cellulolytic
    anaerobes may yield cellulase preparations which are superior to those obtained
    from aerobic species. These organisms also offer the advantage of potentially
    lower enzyme production costs.

2. The specific activity of cellulase produced by bacterial anaerobes such as
    Acetivibrio cellulolyticus is high compared to that obtained from aerobic
    sources (bacterial and fungal). This advantage is counterbalanced by the fact
    that the amount of enzyme produced is small.

3. There are some fundamental differences in the biochemistry of cellulose
    hydrolysis by aerobic and anaerobic systems. These include a requirement for
    reduced conditions and for Ca^{++} by cellulases from anaerobes - something not
    observed in aerobic systems.

4. Streptomyces flavogriseus produces cellulase at levels similar to those reported
    for wild-type Trichoderma strains but considerably lower than those obtained
    with genetically improved strains. One possible advantage of S. flavogriseus
    system is the simultaneous production of valuable by-products such as glucose
    isomerase.

REFERENCES

Bayer,E.A., R. Kenig and R. Lamed. 1983. Adherence of Clostridium thermocellum to cellulose. J. Bacteriol. 156: 818-827.

Johnson, E.A. and A.L. Demain. 1984. Probable involvement of sulfhydryl groups and a metal as essential components of cellulase of Clostridium thermocellum. Arch. Microbiol. 137: 135-138.

Johnson, E.A., M. Sakajuh, G. Halliwell, A. Madia and A.L. Demain. 1982. Saccharification of complex cellulosic substrates by the cellulase system of Clostridium thermocellum. Appl. Environ. Microbiol. 43: 1125-1132.

Lamed, R., E. Setter and E.A. Bayer. 1983. Characterization of a cellulose-binding cellulase containing complex in Clostridium thermocellum. J. Bacteriol. 156: 828-836.

MacKenzie, C.R. and D. Bilous. 1982. Location and kinetic properties of the cellulase system of Acetivibrio cellulolyticus. Can. J. Microbiol. 28: 1158-1164.

MacKenzie, C.R., D. Bilouş and K.G. Johnson. 1984. Purification and characterization of an exoglucanase from Streptomyces flavogriseus. Can. J. Microbiol. 30: 1171-1178.

MacKenzie, C.R., D. Bilous and K.G. Johnson. 1984b. Streptomyces flavogriseus cellulase: Evaluation under various hydrolysis conditions. Biotechnol. Bioeng 26: 590-594.

MacKenzie, C.R. and R.E. Williams. 1984. Detection of cellulase and xylanase activity in isoelectric focused gels using plastic - film supported agar substrate gels. Can. J. Microbiol. in press.

Patel, G.B. and C.R. MacKenzie. 1982. Metabolism of Acetivibrio cellulolyticus during optimized growth on glucose, cellobiose and cellulose. Eur. J. Appl. Microbiol. Biotechnol. 16: 212-218.

Factors affecting cellulase production and the
efficiency of cellulose hydrolysis

J.N. Saddler, C.M. Hogan, G. Louis-Seize and E.K.C. Yu
Biotechnology and Chemistry Department
Forintek Canada Corp.
800 Montreal Road
Ottawa, Ontario. K1G 3Z5

ABSTRACT

The cellulase production step has been shown to be the most expensive step
in any of the processes using enzymatic hydrolysis which propose making ethanol
and other liquid fuels from lignocellulosic residues. As Forintek has one of
the world's largest collections of wood decay microorganisms an initial
screening was carried out to see if any species or strains were more active than
the mutated strains of _Trichoderma reesei_ which are used by most other groups.
It was apparent that numerous factors affected the amount of cellulase produced
and the efficiency at which they hydrolysed various pretreated wood substrates.
Culture filtrates of various cellulolytic bacteria and fungi were compared to
see if the various cellulase assays were representative of the filtrates' overall
ability to hydrolyse a realistic substrate, such as steam-exploded wood, to
glucose. It was apparent that overall hydrolytic activity, as measured by
filter paper activity, was not representative of the cellulase complexes' ability
to hydrolyse pretreated lignocellulosic substrates to glucose. The method of
pretreatment and the nature of the substrate was shown to influence how
readily the cellulose and hemicellulose components were hydrolysed. The efficiency
of hydrolysis could also be greatly enhanced if it was combined with simultaneous
fermentation of the liberated sugars to ethanol. This not only removed end
product inhibition of the cellulase enzyme but also greatly reduced the incidence
of contamination because of the immediate removal of the glucose.

INTRODUCTION

During the last six years the Biotechnology and Chemistry programme at
Forintek has looked at the overall process of converting wood residues to higher
value - added products. Although numerous products have been identified (Figure 1)
the products that we have been most interested in are fermentable sugars and
"reactive" lignin. Our assumption was that once fermentable sugars were obtained,
primarily glucose and xylose, we could then use established fermentation processes
which make or have made various products such as ethanol, butanol, acetone, etc.,
and which used other sources of sugar as the substrate. As the production of
ethanol from corn and sugar cane appears to be commercially viable it would seem
that a similar process using less expensive lignocellulosic substrates would
be a realistic alternative providing the cost of producing the sugars was competitive.
As any biomass conversion process will probably always be aimed at producing lower
value, higher volume products we have restricted the range of products to fermentable
sugars and lignin with the assumption that the sugars would then be converted to
large bulk products such as liquid fuels and chemicals. Although our past work
has emphasized the use of the lignin fraction as an adhesive in composite
board construction (Calvé and Shields, 1982) we have estimated that two or
three commercial "biomass conversion" plants would supply enough material for this
market. We are currently characterizing the different types of bioconversion
lignin and evaluating potential markets for this material.

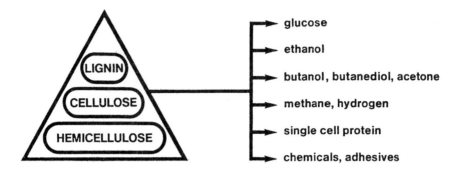

Fig. 1. Potential products that can be obtained by the enzymatic hydrolysis
of pretreated lignocellulosic residues.

During our research on the conversion of wood residues to liquid fuels it
soon became apparent that the major rate-limiting step in the overall process
was the production of the cellulase enzymes and the subsequent hydrolysis of
the various lignocellulosic substrates to glucose. This is also reflected in
the fact that enzyme production costs have been calculated (Wilke et al, 1976)
to account for as much as 60% of the total processing costs associated with the
enzymatic hydrolysis of cellulose to glucose. The low efficiency of hydrolysis
is a multicomponent problem which can be influenced by factors as diverse as the
age of the mycelial inoculum, the nature of the substrate and the configuration
of the vessel in which the fungus is grown. Despite significant advances in
increasing the productivity and amount of cellulases that can be produced by
different cellulolytic microorganisms, using mutation and innovative fermentation
techniques (Gallo et al, 1978; Hendy et al, 1982) little progress has been made
towards increasing the specific activity of the cellulases. In this paper we have
described some of the approaches we have taken to increase the efficiency of
enzymatic hydrolysis in an attempt to make the production of fermentable sugars
from lignocellulosic residues competitive with the production of sugars from
other substrates.

MATERIALS AND METHODS

Microorganisms All of the fungi were taken from the Forintek Culture Collection
(Saddler, 1982). A mycelial inoculum was used to initiate growth in shake flasks
containing 150 mL of Vogel's medium (Montenecourt and Eveleigh, 1977) along with
a designated amount of Solka floc BW 300, glucose or lignocellulosic material.
Fungal cultures were harvested when peak cellulase or xylanase activity was
reached by filtration through a Whatman glass fiber filter. In some instances the
culture filtrates were concentrated by ultrafiltration using the millipore pellicon
cassette system with a molecular weight exclusion limit of 10,000 Daltons.

Saccharomyces cerevisiae C495 (NRCC 202001) was grown under standard conditions
(Rainbow, 1970). Zymomonas mobilis (ATCC 29191) was grown in 60 mL serum vials
(Miller and Wolin, 1974) using the medium of Rogers et al (1979). Clostridium
thermocellum (ATCC 31924 and NRCC 2688) were grown under previously described
conditions (Saddler and Chan, 1982, 1984).

Substrates   Aspen wood chips were saturated with steam at 500 psi for 60 sec., then steam exploded (SEA). Some of the pretreated residue was further extracted at room temperature for 2 hours with water (SEA-WI). A portion of this material was subsequently treated for 1 hour with 0.4% NaOH, washed thoroughly with water, mildly acidified with dilute H_2SO_4 and again washed until the samples were neutral (SEA-WIA). The abbreviations respectively stand for steam-exploded aspenwood (SEA), steam-exploded aspenwood, water insolubles (SEA-WI), steam-exploded aspenwood water insolubles, alkali extracted (SEA-WIA). This procedure and the composition of the pretreated substrates have been described previously (Saddler et al, 1982a, Saddler and Brownell, 1982). Solka floc was purchased from Brown and Co., Berlin, N.H., USA. Larchwood xylan was obtained from Sigma.

Assays. Soluble protein was determined directly after TCA precipitation using the modified Lowry method (Lowry et al, 1981) of Tan et al (1984) and bovine serum albumin (Sigma) as the standard. Total reducing sugars were estimated colorimetrically using dinitrosalicylic acid reagent (Miller, 1959). Glucose was determined colorimetrically by the glucostat enzyme assay (Raabo and Terkildsen, 1960).

Endoglucanase activity (1,4-β-D-glucan 4-glucanohydrolase, EC 3.2.1.4) was determined by incubating 1 mL of culture supernate with 10 mg of carboxymethyl cellulose (Sigma; medium viscosity, D.P. 1100, D.S. 0.7) in 1 mL of 0.05 M sodium citrate buffer, pH 4.8 at 50°C for 30 min. The reaction was terminated by the addition of 3 mL of dinitrosalicylic acid reagent. The tubes were placed in a boiling water bath for 5 min. then cooled to room temperature and the absorbance read at 575 nm.

Filter paper activity was determined by the method of Mandels et al (1976). One mL of culture supernate was added to 1 mL of 0.05 M citrate buffer, pH 4.8, containing a 1 cm x 6 cm strip (50 mg) of Whatman No. 1 filter paper. After incubating for 1 hr at 50°C the reaction was terminated by the addition of 3 mL of dinitrosalicylic acid reagent.

β-glucosidase activity (EC 3.2.1.21) was determined by incubating 1 mL of culture supernate with 10 mg of salicin (Sigma) in 1 mL of 0.05 M citrate buffer, pH 4.8, at 50°C for 30 min. The procedure was the same as for the endoglucanase assay.

Xylanase activity was determined by incubating 1 mL of culture supernate with 10 mg of larchwood xylan (Sigma) in 1 mL of 0.05 M citrate buffer, pH 4.8, at 50°C for 30 minutes. The procedure followed was the same as for the endoglucanase assay.

One unit of activity was defined as 1 μmol of glucose or xylose equivalents released per minute.

Combined hydrolysis and fermentation procedure: 0.5 grams of the cellulosic substrate was added to a 60 mL serum vial and 5 mLs of distilled water. The vials were then gassed with N_2, capped and autoclaved at 15 psi for 20 minutes. Fifteen millilitres of a predetermined cellulase preparation was added aseptically to initiate the experiment and the vials were incubated at 45°C on an orbital shaker (100 rpm) for 24 hrs. One millilitre of a 3 day old S. cerevisiae or Z. mobilis culture was added aseptically after this time and the vials were reincubated at 37°C for various times.

86

Anaerobic culture conditions: The cultures were grown anaerobically at 62°C in
10 mL amounts of DSM medium in capped serum vials (Saddler and Chan 1982, 1984).
C. thermocellum was routinely maintained in 10 mg/mL of solka floc while C. thermohydrosulphuricum and C. thermosaccharolyticum were maintained on 10 mg/mL of
xylose. The co-culture experiments were initiated by the simultaneous inoculation
of equal volumes of each species from 3 day old cultures into DSM medium that
contained the indicated substrate. The inoculum size was 5% by volume of each
species.

RESULTS AND DISCUSSION

     Although numerous microorganisms can degrade soluble derivatives such as
carboxymethyl cellulose, only a comparatively few fungi can produce high levels
of extracellular cellulases capable of extensively degrading insoluble
cellulose to soluble sugars *in vitro* (Mandels, 1975). This is what is required
however if a process using enzymatically hydrolysed lignocellulose to obtain
free sugars is to be economically attractive. To achieve complete hydrolysis
of the insoluble cellulose, the different enzyme components of the cellulase
enzyme complex must be present in the right amounts under the right conditions.
The various cellulase enzyme components and the way in which they are influenced
by the accumulation of the different products are indicated in figure 2. This

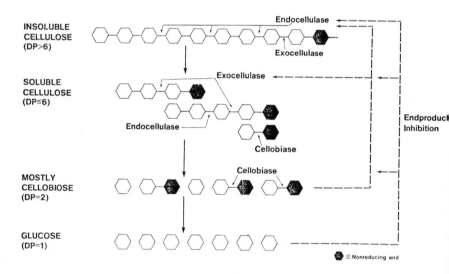

Fig. 2. Schematic representation of the enzymatic hydrolysis of cellulose
        to glucose

requirement for the synergistic action of different endo- and exo-β-D-glucanases
(cellulases (1,4-(1,3;1,4)-β-D-glucan 4-glucanohydrolase, EC 3.2.1.4)) and β-D-
glucosidases (β-D-glucoside glucohydrolase, EC 3.2.1.4) enzymes for complete
cellulose hydrolysis to take place has been reported previously (Wood, 1975;
Ryu and Mandels, 1980; Saddler, 1982). It is this requirement for an active

cellulase complex acting at optimum conditions, combined with the wide range of cellulose substrates on which it must be able to act, which makes it difficult to genetically modify or isolate an ideal cellulolytic microorganism. Some of the factors which can influence cellulase production and the subsequent enzymatic hydrolysis of lignocellulosic substrates are listed in Table 1. Most of the points in the left hand column are process problems which are encountered in any large scale fermentation operation, however several of these points have to be fully resolved before the production of cellulase enzymes can be considered to be economically viable. For example, the majority of the work on cellulase production has used a carefully defined medium which would be uneconomic to reproduce on a larger scale. Similarly, most large scale fermentation processes which produce lower value products such as ethanol tend to be carried out in a shorter period of time as compared to the 4 to 8 days it takes to obtain maximum cellulase yields. Although significant advances have been made at increasing the productivity of the cellulase yields (Ryu and Mandels, 1980; Warzwada et al, 1983) and in the use of continuous and semi-batch cultivation techniques (Hendy et al, 1982) this is an area which still needs to be improved.

Table 1. Factors which influence cellulase production and the efficiency of enzymatic hydrolysis

. Type of inoculum	. Release of enzymes (extracellular)
. Nature of the substrate	. Full spectrum of activities
. Concentration of the substrate	. Constitutive, no feedback inhibition
. Composition of the media	. Half life at elevative temperatures
. Effect of pH	. Enzyme stability over lengthy storage
. Duration of incubation	. Specific activity

The main areas of concern however are those related to the properties of the cellulase enzyme system. As cellulose is a relatively complex, insoluble substrate an extracellular, synergistic enzyme system is required to completely hydrolyse it to glucose. Thus an ideal cellulase complex would have all the components identified in figure 2 which were able to synergistically attack various cellulosic substrates which differed in their degree of polymerisation and degree of crystallinity. So far, no true constitutive cellulase producers have been identified among the cellulolytic fungi. Even the hyperproducing cellulase strains of T. reesei which can grow on lactose still require a cellulosic substrate to induce higher levels of enzyme production (Vandecasteele and Pourquie, 1984). Several of the bacteria however can produce cellulases when they are grown on glucose alone and one of the most cellulolytic has been shown to be Clostridium thermocellum (Wang et al, 1983; Saddler and Chan, 1982).

Although a great deal of work has concentrated on the direct conversion of cellulose to ethanol using C. thermocellum this has not yet been proven to be commercially viable. The major problems are, the organisms inability to grow at high substrate concentrations, its inability to utilise the hemicellulose derived sugars and the relatively low final ethanol concentrations obtained. Although different attempts have been made at circumventing these problems the most successful approaches have involved co-culturing C. thermocellum with non-cellulolytic thermophilic strains, C. thermohydrosulphuricum or C. thermosaccharolyticum (figure 3), which can ferment both the hemicellulose and cellulose derived sugars to ethanol (Ng et al, 1981; Saddler and Chan, 1984).

This resulted in complete utilization of the available sugars and increased ethanol production because of the non-cellulolytic anaerobe's ability to produce higher concentrations of ethanol. Currently various groups are studying different mutation and fermentation methods to enhance ethanol production and substrate utilization by different cellulolytic bacteria.

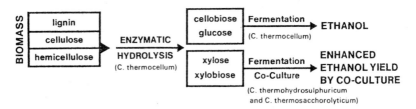

Fig. 3. Diagramatic representation of direct anaerobic conversion of lignocellulosic residues to ethanol using C. thermocellum in mono- and co-culture.

A previous screening of Forintek's culture collection indicated that there were some naturally occurring strains of cellulolytic fungi which appeared to be as hydrolytic as the various T. reesei mutants (Saddler, 1982). We compared the various enzyme activities of T. reesei Rut C30 with one of the strains from our culture collection, T. harzianum E58, which was known to be highly hydrolytic (Table 2). Neither of the strains appeared to be constitutive producers of cellulase confirming the previous observation that T. reesei C30 was more representative of a derepressed mutant (Montenecourt and Eveleigh, 1979). T. reesei C30 did produce considerably more extracellular protein indicating that the high hydrolytic activity of the culture filtrates was probably due to more enzyme being present. When the culture filtrates from the two fungi were used to hydrolyse a range of pretreated wood substrates (Table 3) comparable reducing sugars values were obtained. There was a major difference however in the proportion of glucose that was obtained. The higher β-glucosidase activity of T. harzianum culture filtrates was reflected by the fact that the majority of the sugar was detected as glucose.

Table 2. Cellulase activity of Trichoderma harzianum E58 and T. reesei C30 culture filtrates after 5 and 8 days growth respectively on 2% solka floc and glucose

Trichoderma strain	Substrate (20 mg/mL)	Protein (mg/mL)	Specific activity (IU/mg)		
			Endoglucanase	β-glucosidase	Filter paper
T. harzianum E 58	Solka floc	1.5	33.1	1.8	2.2
	Glucose	0.6	3.6	0.4	0.3
T. reesei C30	Solka floc	3.3	15.4	0.2	1.7
	Glucose	1.0	9.9	0.1	0.6

Table 3. Hydrolysis of various cellulosic substrates by culture filtrates of T. harzianum E58 and T. reesei C30[a] grown on 1% solka floc

cellulase source	Substrates[b] (50 mg/mL cellulose equivalents)	reducing sugars released (mg/mL)		glucose released (mg/mL)	
		24 hrs	120 hrs	24 hrs	120 hrs
T. harzianum E58	Solka floc	15.9	26.5	13.8	26.2
	SEA	10.4	19.9	8.3	17.7
	SE-WI	13.6	23.8	11.4	24.8
	SE-WIA	14.3	25.4	12.8	25.0
T. reesei C30	Solka floc	13.7	24.4	7.0	12.1
	SEA	9.8	19.1	4.7	10.3
	SE-WI	11.2	21.2	6.4	12.0
	SE-WIA	13.4	23.8	7.6	14.6

Culture filtrates from a 5 day old culture of T. harzianym E58 (protein conc. of 1 mg/mL) and 8 day old culture of T. reesei C30 (protein conc. of 3.1 mg/mL) were incubated at 45°C with the indicated substrates

See Materials and Methods for full description of substrates

All of the enzyme components of the cellulase enzyme system are susceptible to product-inhibition by cellobiose and, or glucose (figure 2). One method which we (Saddler et al,1982b, Saddler and Mes-Hartree, 1983) and other workers (Emert and Katzen, 1980) have used to circumvent this problem is to combine the enzymatic hydrolysis with the fermentation step in a combined hydrolysis and fermentation (CHF) process. In this way product inhibition is reduced by directly converting the liberated glucose to ethanol while the incidence of contamination should be reduced by the immediate removal of the glucose. When various steam exploded wood fractions were used as substrates for the CHF process (Table 4) the fractions which had been subsequently extracted by water or alkali were as readily converted to ethanol as was the solka floc control, once the actual cellulose content of the substrate was taken into consideration. The unextracted substrate contained inhibitory materials, presumably decomposition products from the lignin and hemicellulose components, which restricted enzymatic hydrolysis and subsequent fermentation of the liberated sugars (Mes-Hartree and Saddler, 1983; Saddler et al, 1982b). The inhibitors could be removed by extracting the steam exploded substrate with water. This also extracted about 70% of the hemicellulose component which could be used as the substrate for hydrolysis and conversion to butanol or utanediol (Yu et al, 1984).

Although we were able to increase the efficiency of hydrolysis compared to that achieved with some of the earlier T. reesei mutants we have not been able to substantially increase the specific activity of the cellulases. We have used the term specific activity here to mean the amount of extracellular protein required to hydrolyse a stated amount of cellulose. Previous attempts t increasing the efficiency of cellulose hydrolysis have primarily concerned themselves with increasing the productivity of the strains or at increasing the amount of

Table 4. The combined enzymatic hydrolysis and fermentation of pretreated aspenwood[a]

Substrates[b] (5% cellulose equivalents)	Fermentative microorganisms	Reducing sugar (mg/mL)	Glucose (mg/mL)	Ethanol (mg/mL)	% conversion of cellulose to ethanol[c]
Solka floc	None	38.2	31.0	-	-
	S. cerevisiae	1.5	< 0.1	22.1	77.7
	Z. mobilis	0.1	0.5	23.3	81.9
SEA	None	20.5	3.9	-	-
	S. cerevisiae	18.6	1.3	4.8	30.9
	Z. mobilis	20.6	3.2	3.3	21.2
SEA-WI	None	25.8	23.8	-	-
	S. cerevisiae	1.3	<0.1	16.8	86.5
	Z. mobilis	1.5	<0.1	17.0	87.5
SEA-WIA	None	26.8	23.6	-	-
	S. cerevisiae	1.7	<0.1	21.5	83.0
	Z. mobilis	1.2	0.1	22.3	86.1

[a] Hydrolysis was carried out at 45°C for 24 hrs using cellulase enzymes from T. harzianum E58 at an activity of 40 IU Filter paper/g dry wt substrate. Combined enzymatic hydrolysis and fermentation was carried out at 37°C for 72 hrs. This procedure has been previously described (Saddler et al, 1982b).

[b] See Materials and Methods for full description of substrates

[c] % conversion of cellulose to ethanol was calculated as described previously (Saddler and Mes-Hartree, 1983)

extracellular protein (Table 5). Although these higher values have significantly increased the efficiency of hydrolysis it is unlikely that these very high levels of extracellular protein can be increased much more than the levels of approxima* 40 mg/mL reported by some workers (Vandecasteele and Pourquie, 1984).

Cellulases are known to be far less efficient at hydrolysing cellulose as amylases are at hydrolysing a corresponding amount of starch. Both starch and cellulosic conversion processes have many steps in common with the raw material requiring preparation or pretreatment, hydrolysis, fermentation, recovery and concentration of the alcohol and recovery of by-products. The major differences between the two processes can be attributed to both the complexity of the substrate and the enzyme system required to hydrolyse cellulosic substrate (Table 6). Most pretreatment methods either reduce the crystallinity of the cellulose or help dissociate it from the lignin. Even if pure cellulose is used as the substrate the hydrolysis reaction either proceeds for a relatively long period of time, requires substantial amounts of enzyme or results in incomplete hydrolysis of the substrate. Before enzymatic hydrolysis of cellulose to glucose can be expected to be competitive with other sources of sugar it is important that we fully understand the nature of the synergistic action of cellulase hydrolysis and, more importantly, substantially increase the specific activity of the cellulase enzymes.

Table 5. Cellulase production by mutant strains of $\underline{T}$. $\underline{reesei}$

Strain	Soluble Protein (mg/mL)	Specific Activity (IU/mg)			Productivity FPA (IU/L/h)
		CMC	Filter paper	β-glucosidase	
QM 6a[a]	7	12.6	0.7	0.04	15
QM 9414[a]	14	7.8	0.7	0.04	30
NG 14[a]	21	6.4	0.7	0.03	45
C 30[b]	14	8.0	0.7	0.35	68
MCG 77[b]	10	6.5	0.8	0.06	105
RL-P37[c]	8	46.8	1.3	0.25	108
CL-847[b]	22	ND	0.8	0.64	125

[a] data from Ryu and Mandels (1980)

[b] data from Warzywoda et al (1983)

[c] data from Sheir-Neiss and Montenecourt (1984)

ND - not determined

Table 6. Major deterrents to the efficient enzymatic hydrolysis of lignocellulosic residues

. association of the cellulose and hemicellulose with lignin
. crystallinity of the cellulose component
. need for synergistic action of cellulase complex
. low specific activity of the enzymes

REFERENCES

Calvé, L. and Shields, J.A. (1982) Bioconversion of lignins as waferboard and plywood adhesives. Fourth Bioenergy R & D Seminar, Winnipeg, March pp 403-407. NRCC publication 20414

Emert, G.H. and Katzen, R. (1980) Chemtech, October pp 610-614

Lowry, O.H., Rosebrough, J.N., Farr, A.L. and Tandal, R.J. (1951) J. Biol. Chem. 193 265-275

Mandels, M. (1975) in Biotech. Bioeng. Symp. No. 5 (C.R. Wilke, ed) John Wiley and Sons, New York, p. 81

Mandels, M., Andreotti, R. and Roche, C. (1976) in Biotechnol. Bioeng. Symp. No. 6 p. 21-34

Mes-Hartree, M. and Saddler, J.N. (1983) Biotechnol. Lett, 5, 531-536

Miller, G.L. (1959) Anal Chem 31, 426-428

Miller, T.L. and Wolin, M.J. (1974) Appl. Environ Microbiol 27, 985-987

Montenecourt, B.S. and Eveleigh, D.E. (1979) Adv. Chem. Ser, 181, 289

Ng, T.K., Ben-Bassat, A. and Zeikus, J.G. (1981) Appl. Environ. Microbiol., 41, 1337

Rogers, P.L., Lee, K.J. and Tribe, D.E. (1979) Biotechnol Lett. 1 165-170

Ryu, D.D.Y and Mandels, M. (1980) Enzyme Microbial Technol. 2 91

Saddler, J.N. (1982) Enzyme Microbial Technol. 4, 414-418

Saddler, J.N. and Chan, M.K-H (1982) Eur. J. Appl Microbiol Biotechnol 16, 99-104

Saddler, J.N., Brownell, H.H., Clermont, L.P. and Levitin, N. (1982a) Biotechno Bioeng 24, 1389-1402

Saddler, J.N., Hogan, C., Chan, M.K-H and Louis-Seize, G.. (1982b) Can. J. Microbiol 28, 1311-1319

Saddler, J.N. and Brownell, H.H. (1983) in Proc. Roy. Soc. Can. Symp. on Ethano from Biomass, ed H.E. Duckworth, Winnipeg, pp 206-230

Saddler, J.N. and Chan, M.K-H. (1984) Can. J. Microbiol 30, 212

Saddler, J.N. and Mes-Hartree, M. (1984) proc 3rd Pan American Symposium on Fuels and Chemicals by Fermentation, Antigua pp. 104-137

Tan, L.U.L., Chan, M.K-H. and Saddler, J.N. (1984) Biotechnol Letts 6, 199-204

Vandecasteele, J.P. and Pourquie, J. (1984). Proc VI Inter Symp. on Alcohol Fuels Technol Ottawa, vol. 2 p. 227-233

Wang, D.I.C., Avgerninos, G.C., Biococ, I., Wang, S-D. and Fang, H-Y (1983) Philos. Trans. R. Soc. London Ser. B 300, 323-333

Wilke, C.R., Yang, R.D. and Von Stockar, U. (1976) in Biotech. Bioeng. Symp. 6 (Gaden, E.L., Mandels, M.H., Reese, E.T. and Spano, L.A. eds) John Wiley and Sons, New York pp 155

Wood, T.M. (1975) in Biotech. Bioeng. Symp. No. 5 (C.R. Wilke, ed) John Wiley and Sons, New York, p 111.

Yu, E.K.C., Levitin, N. and Saddler, J.N. (1984) in Dev. Indust Microbiol 25 613-620

CELLULASE FROM AN ACIDOPHILIC FUNGUS

Anh LeDuy
Department of Chemical Engineering, Laval University
Sainte-Foy, Québec G1K 7P4

ABSTRACT

A newly isolated acidophilic cellulolytic fungus (deposited as Patent Culture ATCC-20677) is used for the fermentative production of extracellular cellulolytic enzymes under nonaseptic and nonsterile conditions. The fermentation process is carried out under very acidic conditions of pH 2. Under such conditions, microbial contamination is avoided or minimized to an insignificant level for laboratory scale experiments up to 12 L of culture medium in 5-gallon carboys for a period of 10 days. The culture medium used for enzyme production consists of a carbon source from cellulosics or lignocellulosics (such as Na-CMC, xylan, Avicel cellulose, cellulose powdwer, α-cellulose, newsprints, sawdust, etc. or a mixture of the forementioned), together with simple ingredients, such as $(NH_4)_2SO_4$, K_2HPO_4, $MgSO_4$, and $NaNO_3$. The fermentation is carried out at room temperature (28-30 °C), under aerobic conditions and without controlling the pH. The CMCase and xylanase produced are stable under very simple storage conditions, such as in the fresh culture medium not containing the substrate for a period of 3 days, at any temperature from 0 to 30 °C. These extracellular enzymes have an optimum pH around 3, with the best range of pH from 2.0 to 3.6, for any temperature between 15 and 60 °C. The optimum temperatures are 55 °C for CMCase activity and 25 to 50 °C for xylanase activity, at any pH between 2.0 and 5.2. The filter paper (FPase) volumetric activity (IU) and specific activity (IU/mg protein) of the Trichoderma reesei Rut C-30 culture filtrate (which was prepared in shaken flasks at the laboratory of Dr. A.W. Khan, NRC, Ottawa) are 100 times and 6 times those of the acidophilic fungus culture filtrate, respectively. Efforts are being concentrated on the improvement of the acidophilic fungus FPase productivity, to the competitive level of T. reesei Rut C-30, by (i) creating new mutants from the wild strain, (ii) searching for economical chemical inducers, and (iii) optimizing the fermentation conditions in terms of oxygen supply. These studies are essential prior to testing the process on the pilot scale.

DEFINITIONS

The term "cellulolytic enzymes" as employed hereinafter designates an enzyme complex capable of depolymerizing and hydrolyzing various cellulosics and lignocellulosics to oligosaccharides and ultimately to monosaccharides such as glucose (from cellulose) and xylose (from hemicellulose). The term "cellulase" as employed hereinafter designates an enzyme complex capable of depolymerizing and hydrolyzing cellulose into cellodextrins cellobiose and glucose. The term "hemicellulase" as employed hereinafter designates an enzyme complex capable of depolymerizing hemicellulose into oligo-, di- and monosaccharides, mostly pentoses. Thus, the cellulolytic enzymes comprise cellulase and hemicellulase.

ASEPTIC AND STERILE CONDITIONS FOR ENZYME PRODUCTION AND FOR ENZYMATIC

HYDROLYSIS

The production of liquid fuels such as ethanol from cellulosics/lignocellulosics via enzymatic hydrolysis has not proven to be economically feasible yet. This is because of the high fraction of cost associated with the enzyme production and saccharification steps.

i) Enzyme production: All known commercial and pilot scale processes,   for the production of extracellular cellulolytic enzymes by fermentation of the defined microorganisms (in contrast to mixed flora of unidentified microorganisms) in liquid medium (in contrast to solid-state substrate fermentation),   have to be carried out under aseptic and sterile conditions using pharmaceutical standards for equipment and, also, for operating procedures. This constraint has resulted in capital investment and operating cost requirements which have been excessively high for the commercial production of cellulolytic enzymes.   The high cost,   associated with the sterile and aseptic fermentation on the industrial level is the major obstacle to the commercial production of cellulolytic enzymes as bulk chemicals or bulk biochemicals for other industrial uses.   Examples of these uses are the processes for conversion of cellulosic feedstocks into fuel-grade or chemical-grade ethanol.   Examples of the pilot-scale processes are the Gulf-Arkansas University process (1,2) and the Iotech-Natick process (3,4).

ii) Enzymatic hydrolysis or saccharification: The presently known cellulolytic enzymes have their optimum activities within the pH range of 5,0 to 7,0. This implies that the saccharification process for the enzymatic depolymerization and hydrolysis of cellulosics and lignocellulosics should be carried out within this pH range, which is subject to all types of microbial contamination from a nonsterile environment.   In order to avoid or to minimize microbial contamination during the saccharification and the storage periods, either sterile conditions have to be applied or antibiotics used. An example is the work of Reese and Mandels (5) who found that tetracycline and chlortetracycline are the best antibiotics for controlling microbial contamination during saccharification using the cellulolytic enzymes of Trichoderma reesei QM 9414.

During the last decade, research efforts were concentrated on the improvement of the enzyme productivity and quality. The productivity/quality of the enzymes have been significantly improved by optimizing the physical/chemical conditions for fermentation; by creating new mutants such as the strains Natick-MCG77, Rut-C30, Cetus-L27 (6,7) from the wild strains; and even by cloning the Trichoderma reesei cellulase genes into the fast growing bacterium Eschericia coli (8,9). However, the requirement of sterile/aseptic operating conditions still remains as an economic disadvantage for large scale production.

SEARCH FOR NONASEPTIC AND NONSTERILE PROCESS

Fermentation processes could be operated under (i) aseptic and sterile conditions, such as the case of all pharmaceutical processes; (ii) sanitary condition, such as the case of dairy and brewing processes; and (iii) nonaseptic and nonsterile conditions, such as the case of biological processes for wastewater treatment.

In the aseptic and sterile process, the fermenter and all equipments, accessories, materials, ingredients which are in direct contact with the culture medium have to be sterilized and germ free. The inoculum used to seed the fermenter has to be a pure culture. All operation conditions have to be aseptic in order to avoid microbial contamination.

In the sanitary process, the fermenter and all equipments, accessories, materials, ingredients which are in direct contact with the culture medium have to be sterilized or disinfected. The inoculum used to seed the fermenter is preferably pure culture but not required to be absolutely pure. All operating conditions are non aseptic.

In the nonaseptic and nonsterile process, the microbial cultures and the fermentation are conducted in an open tank, container or pond. The culture medium requires no sterilization and the operation requires no special precaution against microbial contamination.

Nonaseptic and nonsterile process for axenic or pure cultures does not exist. However, it is possible to develop a nonaseptic and nonsterile process for the cultivation or the fermentation of one single species (or single strain). It is based on the following principle: the selected microorganism used for the production of a desirable fermentation product has to be able to grow and to synthesize the product under extreme physiological conditions of pH, temperature, chemical environments, etc. Under such conditions, the contamination by other species could be eliminated or reduced to an insignificant level.   If the conditions and the microorganism are appropriately choosen , the fermentation could be operated under nonaseptic and nonsterile condition with the desired strain as in the case of the pure culture.   Table 1 lists some axenic cultures potentially useful for the production of extracellular cellulolytic enzymes under nonaseptic and nonsterile conditions.

Table 1 - Axenic cultures potentially useful for the production of extracellular cellulolytic enzymes under nonaseptic and nonsterile conditions.

Microorganism	Enzyme production	Enzyme properties	References
THERMOPHILES:			
Thielavia terrestris	40–80 °C pH 4.5–6.0	Crude CMCase and Avicelase Optimal temperature 65–80 °C Optimal pH 5.5–8.0	10, 11
Thermomonospora sp.	55 °C pH 7.2	Crude CMCase and Avicelase Optimal temperature 70 °C (CMCase) and 65 °C (Avicelase) Optimal pH 5.9 (CMCase) and 7.0 (Avicelase)	12, 13
Clostridium thermocellum	60–64 °C Anaerobic	Crude CMCase and Avicelase Optimal temperature 64–65 °C Optimal pH 5.2–5.4	14
ALKALOPHILES:			
Bacillus sp. N-4	pH 8–11 37 °C	Partially purified CMCase Optimal pH 10 Optimal temperature 50 °C	15
ACIDOPHILES:			
Acidophilic fungus NC-11	pH 2 28–30 °C	Crude CMCase and xylanase Optimal pH 3 Optimal temperature 55 °C (CMCase) and 25–50 °C (xylanase)	16–18

ACIDOPHILIC FUNGUS AND ACIDIC PROCESS FOR THE FERMENTATIVE PRODUCTION OF

CELLULOLYTIC ENZYMES

Recently, we succeeded in developing a novel process for the production of extracellular cellulolytic enzymes by fermentation under nonaseptic and nonsterile conditions. It consists of carrying out this fermentation process under very acid conditions of pH < 3 by using the appropriate strain of acidophilic/acid-resistant/acid-tolerant cellulolytic microorganism. Microbial contamination is thus avoided or minimized to an insignificant level under this acid pH condition. The concept of this novel process has been proven on the laboratory scale, in up to 12L of culture medium in 5-gallon carboys for a period of 10 days, by using the newly isolated acidophilic fungus strain NC-11 from our private (LeDuy) culture collection of acidophilic fungi. This strain NC-11 has been deposited at the American Type Culture Collection under accession number ATCC-20677 for patent application purposes. A patent application has been filed for this process. We believe that the discovery of the acidophilic cellulolytic fungi, together with the invention of the acidic fermentation process will have significant impact on the commercial application due to the nonrequirement of aseptic/sterile conditions for full-scale operations. All scientific and technical details on the process and the products are reported elsewhere (16-18).

The most relevant features of this novel process are the following:

1) This method of fermentative production of extracellular cellulolytic enzymes under nonaseptic and nonsterile conditions requires very low capital investment and operating cost for industrial scale application. Fermentation can be conducted in an open tank, container or pond with plastic lining instead of a very expensive stainless steel pharmaceutical aseptic fermenter. The culture medium requires no sterilization and the operation requires no special precaution against microbial contamination.

2) The acidophilic fungus strain NC-11 grows and produces at least two enzymes of the extracellular cellulolytic enzymes, the carboxymethylcellulase (CMCase) and the xylanase in a culture medium at pH 2.0 under nonsterile and nonaseptic conditions. The filter paper (FPase) volumetric activity (IU) and specific activity (IU/mg protein) of the Trichoderma reesei Rut C-30 culture filtrate (which was prepared in shaken flasks at the laboratory of Dr. A.W. Khan, NRC, Ottawa) are 100 times and 6 times those of the acidophilic fungus culture filtrate, respectively.

3) The culture medium for this production consists of a carbon source from cellulosics/lignocellulosics, such as Na-CMC, xylan, Avicel cellulose, cellulose powder, α - cellulose, sawdust, newsprints, etc. or a mixture of the forementioned, together with simple ingredients, such as $(NH_4)_2SO_2$, K_2HPO_4, and traces of $MgSO_4$ and $NaNO_3$. The pH is adjusted

to 2.0 with a mineral acid such as H_2SO_4. Fermentation is carried out at room temperature, preferably at 28-30 °C, under aerobic conditions and without control of the pH. The two economic advantages are thus accounted for: the requirement of simple and inexpensive ingredients and the lack of need to control the pH due to the self-stabilization of the culture medium pH during the fermentation process.

4) The CMCase and xylanase produced from the acidophilic fungus strain NC-11 are stable under very simple storage conditions, such as in the fresh culture medium without substrate for a period of 3 days at any temperatures from 0 to 30 °C. These extracellular enzymes have an optimum pH around 3.0, with the best range of pH from 2.0 to 3.6, for any temperature between 15 and 60 °C. The optimum temperatures are 55 °C for CMCase activity and 25 to 50 °C for xylanase activity, for any pH between 2.0 and 5.2. Thus, the extracellular cellulolytic enzymes of the acidophilic fungus strain NC-11 are entirely different from other presently known cellulolytic enzymes due to its optimum acidic pH range from 2.0 to 3.6 for enzyme activities.

5) This implies that the saccharification process or the depolymerization and hydrolysis of cellulosics/lignocellulosics can be operated under acidic conditions, such as at pH 2 or 3 for the best enzyme activities at the same time controlling the microbial contamination. Thus, this process offers further economic advantages for industrial applications because the extracellular cellulolytic enzymes can be stored and used under nonaseptic and nonsterile conditions.

6) In addition, when the cellulosics/lignocellulosics are pretreated with any acidic process prior to the saccharification step, neutralization of the pretreated slurries can be avoided or minimized to a very low cost. This can be achieved by utilizing the extracellular cellulolytic enzymes for saccharification, thereby, saccharification proceeds under acid conditions.

Current efforts are being concentrated on the improvement of the acidophilic fungus FPase productivity, to the competitive level of T. reesei Rut C-30, if possible, by (i) creating new mutants from the wild strain NC-11, by (ii) searching for economical chemical inducers, and (iii) optimizing the fermentation conditions in terms of oxygen supply. These studies are essential prior to testing the process on the pilot scale.

ACKNOWLEDGMENT

Among several students/researchers who were or are working with the acidophilic fungi under my supervision, the person who brought the most contribution to this project is Ms. Nicole Cauchon. Most results reported in this paper are derived from her master thesis.

REFERENCES

(1) Emert, G.H., Katzen, R., Fredrickson, R.E., and Kaupisch, K.F. (1980): Economic Update of the Gulf Cellulose Alcohol Process, Chem. Eng. Progress, 76(9): 47-52.

(2) Anonymous (1982): University of Arkansas: The Cellulose-to-Ethanol Process, in Alcohol Fuels Program Technical Review, Solar Energy Research Institute, pp. 57-74.

(3) Chahal, D.S. (1982): Enzymatic Hydrolysis of Cellulose: State-of-the-Art, National Research Council Canada.

(4) Stone, J.E. (1982): Analysis of Ethanol Production Potential from Cellulosic Feedstocks, Energy Mines Resources Canada.

(5) Reese, E.T. and Mandels, M. (1980): Stability of the Cellulase of Trichoderma reesei under Use Conditions, Biotechnol. Bioeng., 22: 323-335.

(6) Montenecourt, B.S., Kelleher, T.J., and Eveleigh, D.E. (1980): Biochemical Nature of Cellulases from Mutants of Trichoderma reesei, Biotechnol. Bioeng. Symp. No. 10, pp. 15-26.

(7) Shoemaker, S., Watt, K., Tsitovsky, G., and Cox, R. (1983): Characterization and Properties of Cellulases Purified from Trichoderma reesei Strain L27, Bio/Technology, 1: 687-690.

(8) Shoemaker, S., Schweickart, V., Ladner, M., Gelfand, D., Kwok, S., Myambo, K., and Innis, M. (1983): Molecular Cloning of Exo-cellobiohydrolase I Derived from Trichoderma reesei Strain L27, Bio/Technology, 1: 691-696.

(9) Teeri, T., Salovuori, I., and Knowles, J. (1983): The Molecular Cloning of the Major Cellulase Gene from Trichoderma reesei, Bio/Technology, 1: 696-699.

(10) Bellamy, W.D. and Chakrabarty, A.M. (1974): Soluble Cellulase Enzyme Production, U.S. Patent No. 3,812,013.

(11) Skinner, W.A. and Tokuyama, F. (1978): Production of Cellulase by a Thermophilic Thielavia terrestris, U.S. Patent No. 4,081,328.

(12) Hagerdal, B., Ferchak, J.D., and Pye, E.K. (1980): Saccharification of Cellulose by the Cellulolytic Enzyme System of Thermomonospora sp. I. Stability of Cellulolytic Activities with Respect to Time, Temperature, and pH, Biotechnol. Bioeng., 22: 1515-1526.

(13) Moreira, A.R., Phillips, J.A., and Humphrey, A.E. (1981): Production of Cellulases by Thermomonospora sp., Biotechnol. Bioeng., 23: 1339-1347.

(14) Ng, T.K., Weimer, P.J., and Zeikus, J.G. (1977): Cellulolytic and Physiological Properties of Clostridium thermocellum, Arch. Microbiol., 114: 1-7.

(15) Horikoshi, K., Nakao, M., Kurono, Y., and Sashihara, N. (1984): Cellulases of an Alkalophilic Bacillus Strain Isolated from Soil, Can. J. Microbiol., 30: 774-779.

(16) Cauchon, N. and A. LeDuy (1984): Effect of Dilution on Carboxymethylcellulase and Xylanase Assays, Biotechnol. Bioeng., 26: 988-991.

(17) Cauchon, N., and A. LeDuy (1984): Novel Process for the Production of Cellulolytic Enzymes, Biotechnol. Bioeng. (accepted for publication on July 31 1984).

(18) Cauchon, N., (1984): Isolement et caractérisation d'un fungus acidiphile cellulolytique: production des enzymes et évaluation partielle des propriétés biochimiques, M.Sc. Thesis, Université Laval.

The characteristics and cloning of bacterial cellulases

Cecil W.Forsberg[1] ,Kim Taylor[1] ,Bill Crosby[2] ,and David Y.Thomas[3]
1- Department of Microbiology, University of Guelph, Guelph
Ontario N1G 2W1.
2- Plant Biotechnology Institute, N.R.C.C., Saskatoon,
Sakatchewan S7N OW9.
3- Biotechnology Research Institute, N.R.C.C., Ottawa,
Ontario K1A OR6.

Introduction

We are interested in cloning the genes for the various cellulases and for xylanases from the rumen anaerobe Bacteroides succinogenes. Micro-organisms from a wide variety of habitats can use cellulose as a carbon source. Cellulolytic bacteria are known from many genera, some of these are in pure culture capable of degrading and utilizing crystalline cellulose, but others do not possess the full complement of cellulase activities. Bacterial cellulase systems are composed of up to three types of enzymes, exoglucanases, endoglucanases, B-glucosidases, and binding factors. These functions are either located at the cell membrane or secreted into the medium where they can access their substrate.

The biochemical fractionation and characterisation of cellulase systems from some organisms has sometimes proved difficult. Among the dificulties encountered are the identification of individual enzymes from amongst a number of similar acivities which may be related to each other but have been variously modified by post-translational events such as non-specific proteolysis in the medium. Additionally the membrane attachment and hydrophobic nature of some of these proteins has often made conventional methods of fractionation unsuitable.

Cloning of the genes for the individual cellulases in a heterologous host such as E.coli offers a simple method of separation of the components of the cellulase system. Additionally the methods now available for the overproduction of proteins in E.coli can be used to enhance their yield. The purified components can then be reconstituted to determine, for instance, their synergistic interaction. These genes can also be expressed in other hosts such as the yeast Saccharomyces cerevisiae, after the insertion of eukaryotic transcriptional regulatory sequences and modification to provide compatibility with the host cell's secretion apparatus.

The components of bacterial cellulase systems

Exoglucanases (cellobiohydrolases) digest cellulose from the non-reducing end to form principally cellobiose. Exoglucanases and endoglucanases are thought to act synergistically to hydrolyse crystalline cellulose. However, in the case of Trichoderma cellobiohydrolase I it has been reported that this enzyme alone is capable of hydrolysing crystalline cellulose (Chanzy et al.,1983).

In bacteria only the activities present in <u>Cellulomonas</u> <u>uda</u> and <u>Microbispora</u> <u>bispora</u> have been shown to be cellobiohydrolases although several species can grow on crystalline cellulose (see Table 1) including <u>Bacteroides succinogenes</u>. An exoglucanase has not been detected in this organism although a soluble p-nitrophenyl cellobiosidase was found (Schellhorn and Forsberg,1984).

Endoglucanases - all cellulolytic microorganisms possess endoglucanases which act by randomly cutting cellulose chains to yield glucose, cellobiose, and cellotriose. In many microorganisms a multiplicity of endoglucanases have been detected and, in <u>Trichoderma</u> for example, each endoglucanase appears to have a distinct substrate specifity and mode of action (Shoemaker and Brown, 1978). This multiplicity has been attributed in <u>Trichoderma</u> to post-translational modification by proteases (Gritzali and Brown, 1979). The endoglucanases of bacteria have proved difficult to separate and purify (Table 1). In <u>Cellulomonas</u> <u>fimi</u> up to 10 components with CM-cellulase activity can be detected on non-denaturing polyacrylamide gels, it has been proposed that these arise from three or four components by proteolysis and deglycosylation. This is proposed to give the diversity of products which are still enzymatically active but which are reduced in their substrate affinity. The homology of these products was shown by using antibodies for <u>C</u>. <u>fimi</u>, and for <u>Streptomyces</u> <u>antibioticus</u> which produces five distinct endoglucanases three of which are immunologically related. Ten endoglucanase activities have been detected in culture fluid from <u>Bacteroides</u> <u>succinogenes</u>, but it is not yet clear how many of these arise from post-translational modification (Groleau and Forsberg, 1982).

B-glucosidase- this enzymatic activity often forms a rate-limiting step in the microbial digestion of cellulose to glucose. These enzymes in bacteria are invariably associated with the cell membrane. Few bacteria, however, have been reported to possess B-glucosidases. <u>C</u>. <u>fimi</u> has two intracellular B-glucosidases one of which may be an aryl-B-D-glucosidase as it was more active with PNPG and another which cleaves cellobiose (Wakarchuk et al., 1984). <u>C</u>. <u>thermocellum</u> has both a cellobiase which cleaves cellobiose to glucose extracellularly and a cellobiose phosphorylase involved in the uptake of cellobiose (Park and Ryu, 1983). <u>B</u>. <u>succinogenes</u> possesses both cellobiase and B-glucosidase activities which are mainly cell-associated although small amounts of enzyme may be found in the culture supernatants (Forsberg and Groleau, 1982).

Binding factor- a key factor in the utilization of insoluble substrates by bacteria is thought to be the adherence of the bacteria to those substrates. <u>C</u>. <u>thermocellum</u> is the only bacterium in which the binding factor has been characterized, it is a complex with a molecular weight of 2.1 million, and has been resolved by SDS-gel electrophoresis into 14 bands (Bayer et al., 1983; Lamed et. al., 1983). This complex possesses a major part of the cellulolytic activity and also is responsible for the binding of the cells to cellulose. <u>B</u>. <u>succinogenes</u> also adheres firmly to cellulose. Whether the binding activity is an intrinsic property of the cellulase enzymes, either collectively or individually, or conferred by other components of the complex, is the type of question that can well be addressed by gene cloning.

Table 1: Characteristics of β-(1,4)-glucanases and β-(1,4)-glycosidases from cellulolytic bacteria.

Bacteria	Glucanases							Glycosidases				References
	Exo-	pH	Temp	Endo-	pH	Temp	Loc.	β-G[1]	Cb[2]	Phos[3]	Loc.	
Cytophaga species	Y[4]	7.0	-	6.2K[5]	8.0	-	PER[6]	Y	-	-	PER	Chang and Thayer, 1977
	Y	-	-	8.7K	-	-	CYTO[7]	-	-	-	-	Marshall, 1972, 1973
Cellvibrio fulvus	Y	-	-	Y	7.0	-	CB CF[8]	-	-	Y[9]	-	Berg et al., 1972
Cellvibrio vulgaris	n.d.[10]	-	-	Y	4.6	37	CF	-	-	Y	-	Oberkotter and Rosenberg, 1978
Cellulomonas CS1-1	n.d.	-	-	Y	-	-	CF	-	-	-	-	Choi et al., 1978
Cellulomonas sp.	-	-	-	51K 54K 118K	-	-	CF	-	-	-	-	Haggett et al., 1978
Cellulomonas	Y	-	-	-	-	-	-	-	-	Y	-	Schimz et al., 1983
Cellulomonas fimi	-	-	-	10	-	-	CF	Y	Y	-	CB	Langsford et al., 1984, Wakarchuk et al., 1984
Cellulomonas uda	66K	5.5	45	-	-	-	CF					Nakamura and Kitamura, 1983
Sporocytophaga myxococcoides	Y	-	-	46K	5.5	-	CB-CF	-	-	-	-	Vance et al., 1980
	-	-	-	52K	7.5	-	-	-	-	-	-	Osmundsvag and Goksoyr, 1975
Pseudomonas fluorescens v. cellulosa	n.d.	-	-	40K-85K	7.0	-	CB-CF	Y	-	-	MEM[11]	Yoshikawa et al., 1974 Yamane et al., 1970
Thermomonospora spp.	4[12]	7.0	65	4	5.9	70	CF	Y	Y	-	CB	Lamed et al., 1983 Ferchak et al., 1980 Hagerdal et al., 1980
Streptomyces flavogriseus	y[13]	5.6	40	Y	7.0	50	CF	Y	Y	-	CB	Ishaque and Kluelpfel, 1980
Rhizobium trifoli	n.d.	-	-	Y	-	-	-	-	-	-	-	Martinez-Molina et al., 1979

(cont'd)

Bacteria	Glucanases							Glycosidases				References
	Exo-	pH	Temp	Endo-	pH	Temp	Loc.	β-G[1]	Cb[2]	Phos[3]	Loc.	
Acetovibrio cellulolyticus	38K	5.0	50	33K 10K	5.0	50	CF	-	81K	-	CF	Saddler et al., 1980. Saddler and Khan, 1981.
Bacteroides cellulosolvens	Y	5.0	40	Y 7[7]	5.0	50	CB-CF	N	N	Y	-	Giuliano and Khan, 1984. Groleau and Forsberg, 1981, 1983
Bacteroides succinogenes	n.d.	-	-	7[7]	6.0	50	CB-CF	Y	Y	n.d.	CB	Gaudet, 1983
Clostridium acetobutylicum	n.d.	-	-	60K Y	4.6	37	CF	-	-	-	CB-CF	Allcock and Wood, 1981, S.F. Lee, unpublished.
Clostridium (M7)	Y	6.5	67	4 Y	6.5	67	CF	Y	Y	-	-	Lee and Blackburn, 1975.
Clostridium polysaccharolyticum	Y	-	-	Y	-	-	-	-	-	-	-	Van Gylswyk et al., 1980.
Clostridium stercorarium	Y	-	-	99	6.4	85	CF	?[15]	?	-	CB	Creuzet and Frixon, 1983, Creuzet et al., 1983.
Clostridium thermocellum	Y	5.4	64	83K- 94K	5.2	62	CF	-	n.d.	-	-	Ng et al., 1977. Ng and Zeikus, 1981b
	Y	5.7	70	56K	6.0	65	CF					Petre et al., 1981.
	Y	6.4	-	Y	6.1	65	CF					Johnson et al., 1982.
	Y[16]	-	-	91K- 99K 816	-	-	CF					Creuzet and Frixon, 1983.
Ruminococcus albus	-	-	-	>620K 620K 49K 39K	-	-	-	-	y[a]	y[b]	CF[a]-CB[b]	Bayer et al., 1983. Lamed et al., 1983. Park and Ryu, 1983.
	Y	-	-	30K	-	-	CF	-	-	Y	-	Yu and Hungate, 1979. Leatherwood, 1965.
R. albus 8	Y[17] 85K 102K	-	-	100K	6.8	37	CF	-	-	-	-	Wood et al., 1982. Ohmiya et al., 1982. Stack and Hungate, 1984.
Ruminococcus flavefaciens	89K	6.5	39	89K	6.4	50	CB-CF	Y	n.d.	Y	-	Pettipher and Latham, 1979.

1_ β-G, aryl β-glucosidase activity (measured using o-nitrophenyl β-D-glucoside).

2_ Cb, cellobiase activity.

3_ Phos, cellobiose phosphorylase.

4_ Y, indicates the presence of the enzyme.

5_ units of molecular weight.

6,7,8,11_ PER, periplasm, CYTO, cytoplasm, CB, cell-bound, CF, cell-free, MEM, membrane.

9_ activity detected in Cellvibrio gilvus (Hulcher and King 1958).

10_ n.d., not detected.

12_ 4, four components present, avicelase activity not separated from CMCase activity.

13_ hydrolysis of avicel not tested but filter paper and cotton readily hydrolysed.

14_ 7, at least 7 separable componentsμ major component with a molecular weight of 43,000.

15_ assay method not indicated, therefore cannot differentiate between activities.

16_ 8 components present in a single 2.1M dalton complex (seven endoglucanases and one non-hydrolytic affinity factor (210K daltons) which binds complex to cellulose).

17_ aryl β-(1,4)-cellobiosidase activity measured using p-nitrophenyl cellobioside.

Cloning of cellulase genes

A cellulase gene from one fungus Trichoderma reesei and several bacterial species has been cloned (Table 2). We have cloned from B.succinogenes a number of different cellulase genes (Crosby et al.,1984) that we are in the process of sequencing and biochemically characterising. The cloning of the bacterial cellulase genes has relied on their expression in E.coli. An immunological method has been used to screen clones for the production of cellulases, and this method does have the advantage of identifying proteins produced in E.coli which are enzymatically inactive (Whittle et al.,1982). We have successfully used the Congo Red polymer binding method developed by Teather and Wood (1982) to screen E.coli transformants which express B.succinogenes cellulases. This technique can detect different cellulase enzymatic activities. Out of 7000 transformants screened by this technique 15 E.coli clones were identified which caused clearing zones on CMC-cellulose plates. Two distinct phenotypes were identified on the basis of the size of the clearing zone, and on the basis of restriction enzyme digestion patterns 6 different genes were identified (Crosby et al.,1984). Subsequent characterisation has shown that for this technique, the enzyme does not have to be truly secreted from E.coli to allow its detection. Few proteins are secreted into the extracellular medium from E.coli, e.g.colicins, others remain in the periplasmic space e.g. B-lactamase. A ts mutant of E.coli has been described which at restrictive temperature leaks periplasmic enzymes, including the cloned C.fimi endoglucanase, into the cell medium(Gilkes et al.,1984).

Table 2: Cellulase genes cloned into Escherichia coli

Organism	Enzyme	Expression and localization *	Refs.
Trichoderma reesei	cellobiohydrolase I	-	Shoemaker et al.1983
Cellulomonas fimi	endoglucanases	largely periplasmic	Gilkes et al. 1984
Clostridium thermocellum	endoglucanases	celA-62% periplasmic celB-30% periplasmic	Cornet et al. 1983
Bacteroides succinogenes	endoglucanases	celA-38% periplasmic celB- 7% periplasmic	Crosby et al. 1984 Taylor et al. 1984
Bacillus sp.	endoglucanases	pNK1-74% periplasmic pNK2-34% periplasmic	Sashihara et al.1984
Thermomonospora sp. YX	endoglucanase	pD316 30% periplasmic 30% extracellular	Collmer and Wilson et al. 1983
Escherichia	B-glucosidase	-	Armentrout and Brown 1981

* Unless otherwise stated the remainder of the enzyme activity was intracellullar.

The endoglucanases cloned in E.coli (Table,2) are mostly secreted into the periplasmic space albeit with varying degrees of efficiency. In C.thermocellum a multiplicity of endoglucanases have been detected but only two genes have been cloned (Table 2) but only two endoglucanase genes have been cloned. The celB endoglucanase in E.coli is detectable as two active bands of 55,000 and 53,000 molecular weight, since about 30% of the celB gene product is found in the periplasmic space this may reflect cleavage of a peptide leader sequence (Cornet et al., 1983). Although why only a fraction of the enzyme is processed is unclear.

We have found that molecular cloning is an efficient method of separating the different cellulase enzymatic activities. We have yet to detect genetically-linked cellulase genes i.e. there does not appear to be an operon for these genes of similar function. The control of the expression of these genes has not been fully examined, the expression of some cellulases seems to be glucose-repressible but enzyme activity is also inhibited by glucose.

In the near future research will concentrate on characterising the cellulase genes. In particular their hyper-expression in a variety of microbial hosts will be examined. For instance a vector expression/secretion system for Saccharomyces cerevisiae has been developed by us (Lolle et al., 1984), and makes possible the development of a cellulolytic yeast.

References

Allcock,E.R. and D.R.Woods. 1981. Carboxymethyl cellulase and cellobiase production by Clostridium acetobutylicum in an industrial fermentation medium. Appl.Environ.Microbiol.41:539-541.

Armentrout,R.W., and R.D.Brown. 1981. Molecular cloning of genes for cellobiose utilization and their expression in Escherichia coli. Appl.Environ.Microbiol. 41:1355-1362.

Bayer,E.A.,R.Kenig,and R.Lamed.1983. Adherence of Clostridium thermocellum to cellulose. J.Bacteriol.156:818-827.

Berg,B.,B.v.Hofsten,and G.Petterson.1972. Growth and cellulase formation by Cellvibrio fulvus. J.Appl.Bact.35:201-214.

Chang,W.T.H., and D.W. Thayer.1977. The cellulase system of Cytophaga species. Can.J.Microbiol.23:1285-1292.

Choi,W.Y.,K.D.Haggett,and N.W.Dunn.1978.Isolation of a cotton wool degrading strain of Cellulomonas: Mutants with altered ability to degrade cotton wool. Aust.J.Biol.Sci.31:553-564.

Collmer,A., and D.B.Wilson. 1983. Cloning and expression of a Thermomonospora YX endocellulase gene in E.coli.Bio/technology1:589.

Cornet,P., J.Millet, P.Beguin, and J.P.Aubert. 1983. Characterization of two cel genes (cellulose degradation) genes of Clostridium thermocellum coding for endoglucanases. Bio/technology 1: 589-594.

Creuzet,N., J.F.Berenger, and C.Frixon.1983.Characterisation of exoglucanase and synergistic hydrolysis of cellulose in Clostridium stercorarium. FEMS Microbil.Lett. 20:347-350.

Creuzet,N., and C.Frixon. 1983. Purification and characterization of an endoglucanase from a newly isolated thermophilic anaerobic bacterium. Biochimie 65:149-156.

Crosby,B., B.Collier, D.Y.Thomas, R.M.Teather, and J.D.Erfle. 1984. Cloning and expression in Escherichia coli of cellulase genes from Bacteroides succinogenes. Biofuels Symp.,Ottawa. Elsevier. in press.

Ferchak,J.D., B.Hagerdal, and E.K.Pye. 1980. Saccharification of cellulose by the cellulolytic enzyme system of Thermomonospora sp. II. Hydrolysis of cellulosic substrates. Bitechnol.Bioeng.22:1527-1542.

Gaudet,G. 1983 Les cellulases extracellulaires de Bacteroides succinogenes. Ann.Microbiol. 134A:111-114.

Gilkes,N.R., D.G.Kilburn, R.C.Miller, and R.A.J.Warren. 1984. A mutant of Escherichia coli that leaks cellulase activity encoded by cloned cellulase genes from Cellulomonas fimi.Bio/technology3:259-263

Giuliano,C., and A.W.Khan. 1984. Cellulase and sugar formation by Bacteroides cellulosolvens, a newly isolated cellulolytic anaerobe. Appl.Environ.Microbiol. 48:446-448.

Gritzali,L., and R.D.Brown. 1979. The cellulase system of Trichoderma Relationships between purified extracellular enzymes from induced and cellulose grown cells. Adv.Chem.ser. 181: 237-260.

Groleau,D., and C.W.Forsberg. 1981. The cellulolytic activity of the rumen bacterium Bacteroides succinogenes. Can.J.Microbiol.27:517-530.

Groleau,D., and C.W.Forsberg. 1983. Partial characterization of the extracellular endoglucanase produced by Bacteroides succinogenes. Can. J. Microbiol. 29: 504-517.

Hagerdal,B., J.D.Ferchak and E.K.Pye. 1980. Saccharification of cellulose by the cellulolytic enzyme system of Thermomonospora sp. I. Stability of cellulolytic activities with respect to time, temperature and pH. Biotechnol.Bioeng. 22: 1515-1526.

Haggett,K.D., W.Y.Choi, and N.W.Dunn. 1978. Mutants of Cellulomonas which produce increased levels of B-glucosidase.Europ.J.Appl. Microbiol.Biotechnol. 6:189-191.

Hulcher,F.H. and K.W.King. 1958. Metabolic basis for disaccharide preference in a Cellvibrio. J.Bacteriol.76:571-577.

Ishaque,M., and D.Kluepfel. 1980. Cellulase complex of a mesophilic Streptomyces strain. Can.J.Microbiol.26:183-189.

Johnson,E.A., M.Sakajoh, G.Halliwell, A.Madia, and A.L.Demain. 1982. Saccharification of complex cellulosic substrates by the cellulase system from Clostridium thermocellum. Appl.Environ.Microbiol.43:1125.

Lamed,R., E.Setter, and E.A.Bayer. 1983. Characterization of a cellulose-binding, cellulase-containing complex in Clostridium thermocellum. J.Bacteriol. 156:828-836.

Langsford,M.L., N.R.Gilkes, W.W.Wakarchuk, D.G.Kilburn, R.C.Miller and R.A.J.Warren. 1984. The cellulase system of Cellulomonas fimi. J.Gen.Microbiol. 130: 1367-1376.

Leatherwood,J.M. 1965. Cellulase from Ruminococcus albus and mixed rumen microorganisms. Appl.Microbiol. 13:771-775.

Lee,B.H. and T.H.Blackburn. 1975. Cellulase production by a thermophilic Clostridium species. Appl.Microbiol. 30:346-353.

Lolle,S., N.Skipper, H.Bussey, and D.Y.Thomas. 1984. The expression of cDNA clones of yeast M1 double-stranded RNA in yeast confers both killer and immunity phenotypes. EMBO Jour. 3:1383-1387.

Marshall,J. 1973. Separation and characterization of the B-D-glucanohydrolase from a species of Cytophaga. Carbohydr.Res. 26: 274-277.

Martinez-Molina,E., V.M.Morales, and D.H.Hubbell. 1979. Hydrolytic enzyme production by Rhizobium. Appl.Environ.Microbiol.38:1186-1188.

Nakamura,K., and K.Kitamura. 1983. Purification and some properties of a cellulase active on crystalline cellulose from Cellulomonas uda. J.Ferment.Technol. 61:379-382.

Ng,T.K., P.J.Weimer, and J.G.Zeikus. 1977. Cellulolytic and physiological properties of Clostridium thermocellum. Arch.Microbiol. 114: 1-7.

Ng.,T.K., and J.G.Zeikus. 1981. Purification and characterization of an endoglucanase (1,4-B-D-glucan glucanohydrolase) from Clostridium thermocellum. Biochem.J. 199:341-350.

Oberkotter,L.V., and F.A.Rosenberg. 1978. Extracellular endo-B-(1,4)-glucanase in Cellvibrio vulgaris. Appl.Environ.Microbiol.36:205-209.

Ohmiya,K., M.Shimizu, M.Taya, and S.Shimizu. 1982. Purification and properties of cellobiosidase from Ruminococcus albus. J.Bacteriol. 150: 407-409.

Osmundsvag,K., and J.Goksoyr. 1975. Cellulase from Sporocytophaga myxococcoides. Purification and properties. Eur.J.Biochem.57:405-409.

Park,W.S., and D.D.Y.Ryu. 1983. Cellulolytic activities of Clostridium thermocellum and its carbohydrate metabolism. J.Ferment. Technol. 61:563-571.

Petre,J., R.Longin, and J.Millet. 1981. Purification and properties of an endo-B-1,4-glucanase from Clostridium thermocellum. Biochimie 63: 629-639.

Pettipher,G.L., and M.J.Latham. 1979. Characteristics of enzymes produced by Ruminococcus flavefaciens which degrade plant cell walls. J.Gen.Microbiol. 110: 21-27.

Saddler,J.N., and A.W.Khan. 1981. Cellulolytic enzyme system of Acetivibrio cellulolyticus. J.Gen.Microbiol. 27:288-194.

Saddler,J.N., A.W.Khan, and S.M.Martin. 1980. Regulation of cellulase synthesis in Acetivibrio cellulolyticus. Microbios. 28: 97-106.

Sashihara,N., T.Kudo and K.Horikoshi. 1984. Molecular cloning of cellulase genes of alkophilic Bacillus sp. strain N-4 in Escherichia coli. J.Bacteriol. 158: 503-506.

Schellhorn,H.E., and C.W.Forsberg. 1984. Multiplicity of extracellular B-(1,4)-endoglucanases of Bacteroides succinogenes S85. Can.J.Microbiol. 30:930-937.

Schellhorn,H.E., and C.W.Forsberg. 1984. Multiple endoglucanase activities of Bacteroides succinogenes. Can.J.Microbiol.,in press.

Schimz, K.L., B.Broll, and B.John. 1983. Cellobiose phosphorylase (E.C.2.4.1.20) of Cellulomonas: occurrence, induction and its role in cellulose metabolism. Arch.Microbiol. 135:241-249.

Shoemaker,S., V.Schweickart, M.Lakmer, D.Gelfand, S.Kwod, K.Myambo and M.Innis. 1983. Molecular cloning of exocellobiohydrolase I derived from Trichoderma reesei strain L27. Bio/technology 1:691-696.

Stack,R.J., and R.E.Hungate. 1984. Effect of 3-phenylpropionic acid on capsule and cellulases of Ruminococcus albus 8. Appl.Environ. Microbiol. 48:218-233.

Taylor,K., B.Crosby, C.W.Forsberg and D.Y.Thomas. 1984. Characterization of the endoglucanase genes from Bacteroides succinogenes in Escherichia coli. in preparation.

Teather,R.M., and P.J.Wood. 1982. Use of Congo red-polysaccharide interactions in enumeration and characterization of cellulolytic bacteria from the bovine rumen. Appl.Environ.Microbiol. 43:777-780.

Vance,I., C.M.Topham, S.L.Blayden, and J.Tampion. 1980. Extracellular cellulase production by Sporocytophaga myxococcoides NCIB 8639. J.Gen.Microbiol. 117: 235-241.

Wakarchuk,W.W., D.G.Kilburn, R.C.Miller, and R.A.J.Warren.1984. The preliminary characterization of the B-glucosidases of Cellulomonas fimi. J.Gen.Microbiol. 180: 1385-1389.

Wood,T.M., and S.I.McCrae. 1979. Synergism between enzymes involved in the solubilization of native cellulose. In, Adv.Chem.Ser.181, pp181-210. eds. R.D.Brown and L.Jurasek.

Wood,T.M. and C.A.Wilson. 1984. Some properties of the endo-(1,4)-B-D-glucanase synthesized by the anaerobic cellulolytic rumen bacterium Ruminococcus albus. Can.J.Microbiol. 30:316-321.

Wood,T.M., C.A.Wilson and C.S.Stewart. 1982. Preparation of the cellulase from the cellulolytic anaerobic rumen bacterium Ruminococcus albus and its release from the bacterial cell wall. Biochem.J. 205:129-137.

Yamane,K., H.Suzuki, and K.Nisizawa. 1970. Purification and properties of extracellular and cell-bound cellulase components of Pseudomonas fluorescens var. cellulosa. J.Biochem. 67:19-35.

Yoshikawa,T., H.Suzuki, and K.Nisizawa. 1974. Biogenesis of multiple cellulase componenets of Pseudomonas fluorescens var.cellulosa. J.Biochem. 75: 531-540.

Yu,I., and R.E.Hungate. 1979. The extracellular cellulases of Ruminococcus albus. Ann.Rech.Vet. 10:251-254.

IN SEARCH OF LIGNIN MODIFYING GENES

Pierre Chartrand[1], Pierre Trudel[1], Danièle Courchesne[1] and Christian Roy[2].
[1]Department of Microbiology, Medical Faculty, [2]Department of Chemical Engineering,
Applied Science Faculty, University of Sherbrooke, Sherbrooke, Quebec.

The aim of this project is to identify and isolate from white rot fungi genes that code for lignin-modifying enzymes.

Lignin-modifying genes are part of secondary metabolism pathways and are generally not expressed in exponentially growing fungi. Lignin itself does not activate these genes. It is the change from active to stationary growth that is responsible for the activation of lignin-modifying enzymes. The passage between the two growth phases can be controlled by modifying the nitrogen concentration of the culture medium. Thus it is possible to achieve differential expression of the lignin-modifying genes. It is also possible to induce certain specific enzymes, such as phenol oxidases, with phenol derivatives.

Differentially expressed genes can be identified by hybridization with radiolabeled mRNA prepared from induced and non-induced cultures. In this way, it is possible to identify genes involved in an inducible metabolic pathway without any knowledge of their function. This was the approach we chose to identify and isolate lignin-modifying genes.

We have prepared a total genomic library of Trametes versicolor, a white rot fungus. We then hybridized this library with cDNA radioactive probes prepared from induced and non-induced mRNA extracts. This enabled us to identify and isolate 49 clones that contain genes that code for mRNAs that are more abundant when the lignolytic activity is present than when it is absent. We have also identified and isolated other clones containing genes that are specifically expressed when laccase, a phenol oxidase, is induced.

This report will present our preliminary results in our efforts to isolate genes coding for enzymes capable of modifying lignin.

Very little is known about enzymatic modification of lignin. The reason for this is that lignin metabolism is part of the secondary metabolism that is activated after the onset of the stationary phase of growth (5). This distinguishes lignin metabolism from the metabolism of most other biological polymers which are associated with the active phase of growth. Other distinctive characteristics are that the lignin-modifying genes are not induced by lignin itself (5) and that the main depolymerization reaction is probably an oxidation, not hydrolysis reaction (1).

Recently, Kirk and collaborators (10-11) and Gold and collaborators (3) have independently isolated from the white rot fungus, Phanerochaete chrysosporium an enzyme called ligninase. This enzyme is an oxygenase that requires peroxide as co-factor. It is capable of oxidizing C_α-C_β intermonomer linkages, commonly found in lignin. Other enzymes are suspected of being related to lignin metabolism, but their role is not well established (4-7-12).

The aim of our project is to clone genes coding for enzymes capable of modifying lignin. Instead of using the standard approach of identifying an enzyme and then trying to clone the corresponding gene, we have chosen a somewhat different approach that would enable us to clone lignin modifying genes without any specific knowledge of their functions. This approach has been used to identify genes involved in cell differentiation (9). A prerequisite to its use is the possibility of differential expression of the genes of interest, i.e. two sets of condition, one in which the genes of interest are highly expressed and another one in which these genes are not expressed, or only weakly expressed.

We present in figure 1, a schematic representation of this approach. First, mRNAs are extracted from induced and non-induced cultures; that is cultures that express or do not express the genes of interest. Both of these mRNA preparations are used to synthesize radioactive cDNA probe. Each probe is hybridized to identical genomic blots. The genomic blots are prepared from a gene library of the microorganism having the desired enzymatic functions. Because the genes of interest are only or much more expressed in the induced cultures, these genes will be visualized only with the induced probe and not with the non-induced probe. This permits one to isolate from the gene library, the genomic segments that are differentially expressed. Each of these segments is purified to homogeneity. In turn this purified genetic segment can be used as a probe to confirm the differential amount of its mRNA in induced versus non-induced cultures.

We have said at the beginning that lignin modifying enzymes were not induced by lignin itself. It has been shown however that low concentration of nitrogen will induce synthesis of lignin-modifying enzyme (6). Low nitrogen concentration will result in stationary phase growth, which activates secondary metabolism. In our hands, we can achieve a ten fold differential expression of lignin-modifying activity by varying the concentration of nitrogen in the medium (figure 2). We could thus proceed with the approach described in figure 1.

The microorganism we have chosen for our study is Trametes versicolor, a white rot fungus. The reason for this choice was not only that Trametes versicolor had a high capacity to modify lignin but also because it produced a relatively well characterized phenol oxidase, known as laccase. This enzyme can be specifically induced with phenol derivatives (8). It would serve as a control to test the feasibility of our approach. Moreover, phenol oxidases have been associated with lignin-modifying activity. In the case of laccase, even though we do not detect its activity in conjunction with the lignin-modifying activity of the low nitrogen cultures, we have found that if we induced laccase activity in low nitrogen culture we find a much increased lignin modifying activity (figure 3). In fact the increase in activity is such that we can not measure its rate accurately. One explanation for these results would be a synergistic effect of laccase and unknown lignin-modifying enzymes.

To proceed with our approach, we have prepared a complete genomic library of Trametes versicolor in a lambda vector. The size of the genome of Trametes versicolor is unknown but from the genome sizes of related microorganisms we have estimated that with 10-15 Kb fragments, we would need 25 to 30,000 phage particles to have a 99% probability of covering the whole genome. We have generated over 250,000 individual phage particles for our gene library.

114

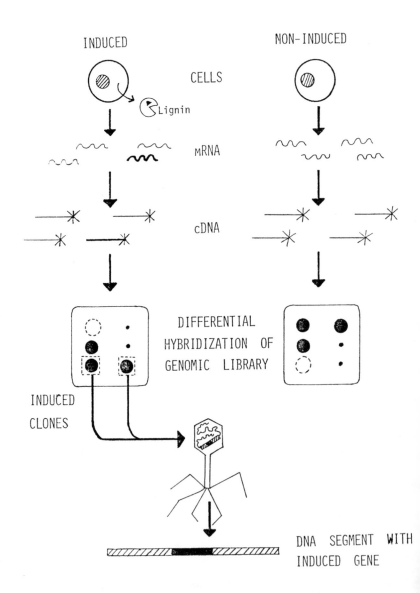

FIGURE 1: Schematic representation of the differential expression approach for cloning genes with unknown functions.

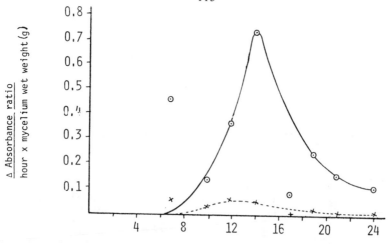

FIGURE 2: Effect of nutrient nitrogen on lignin modifying activity by _Trametes_ _versicolor_. Cultures grown in presence of low (2.4 mM, ⊙) or high nitrogen concentration (33 mM, X). The lignin modifying activity was measured by dye decolorization of poly B411 as described by Glen et al. (2).

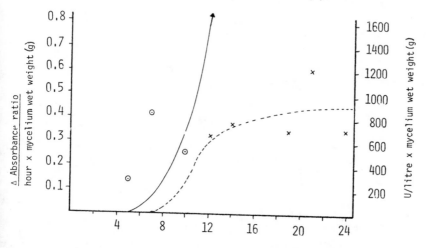

FIGURE 3: Synergistic effect of lignin modifying enzymes and laccase on lignin modifying activity. A culture of _Trametes_ _versicolor_ grown in presence of low nitrogen concentration was induced on the sixth day with xylidine for laccase production; (———) lignin modifying activity, (- - - - - -) laccase activity.

We have screened 50,000 representatives of this library with cDNA probes prepared from Trametes versicolor cultures induced and non-induced for lignin modifying activity. Forty-nine individual plaques that gave the appropriate differential signal were isolated. Four of these have been fully purified and all four have been shown to be different and to code for mRNA that is more abundant in induced than in non-induced cultures.

As a control for this type of approach, using the same technique, we have screened the same bank for the gene coding for the laccase enzyme. This time the induced probe was made from mRNA coming from a culture specifically induced for laccase activity with a phenol derivative. Out of 50,000 representatives of our genomic library 4 plaques gave the appropriate differential signal. Two of these have been fully purified and one of these two, codes for a mRNA only present in cultures with laccase activity. Moreover the molecular weight of this mRNA is the one expected for the mRNA of laccase.

This is where the results stand at the moment. We will continue to characterize the isolated clones, namely by doing physical mapping and comparative hybridization of these clones. We will also do in vitro translation of mRNA corresponding to these clones and study the resulting proteins. These proteins will be used in in vitro assays in which the substrates will be extracted lignin and lignin-models. Ultimately genes of interest will be linked to expression vectors to mass produce the enzymes, for the purpose of modifying lignin into products of interest.

REFERENCES

1- Chang, H-M., Chen, C-L. and T.K. Kirk, 1980. The chemistry of lignin degradation by white-rot fungi. In Lignin biodegradation: Microbiology, Chemistry and Potential Application, vol. 1, p. 215 – 230. Edited by T.K. Kirk, T. Higuchi and H-M. Chang. Florida: CRC Press.

2- Glenn, J.K. and M.H. Gold. 1983. Decolorization of several polymeric dyes by the lignin-degrading basidiomycete Phanerochaete chrysosporium. Appl. Environ. Microbiol. 45, 1741-1747.

3- Glenn, J.K., Morgan, M.A., Mayfield, M.B., Kuwahara, M. and M.H. Gold. 1983. An extracellular H_2O_2-requiring enzyme preparation involved in lignin biodegradation by the white rot basidiomycetes Phanerochaete chrysosporium. Biochem. Biophys. Res. Commun. 114, 1077-1083.

4- Ishihara, T. 1980. The role of laccase in lignin biodegradation. In Lignin biodegradation: Microbiology, Chemistry and Potential Application, vol. 2, p. 17-31. Edited by T.K. Kirk, T. Higuchi and H-M. Chang. Florida: CRC Press.

5- Keyser, P., Kirk, T.K. and J.G. Zeikus. 1978. Ligninolytic enzyme system of Phanerochaete chrysosporium: synthesized in the absence of lignin in response to nitrogen starvation. J. Bacteriol. 135, 790-797.

6- Kirk, T.K., Schultz, E., Connors, W.J., Lorenz, L.F. and J.G. Zeikus. 1978. Influence of culture parameters on lignin metabolism by Phanerochaete chrysosporium. Arch. Microbiol. 117, 277-285.

7- Kuwahara, M., Glenn, J.K., Morgan, M.A. and M.H. Gold. 1984. Separation and characterization of two extracellular H_2O_2-dependent oxidases from ligninolytic cultures of Phanerochaete chrysosporium. FEBS Lett. 169, 247-250.

8- Malmström, G., Fahraeus, G. and R. Mosbach. 1958. Purification of laccase. Biochem. Biophys. Acta 28, 652-653.

9- Mangiarotti, G., Chung, S., Zuker, C. and H.F. Lodish. 1981. Selection and analysis of cloned developmentally-regulated Dictyostelium discoīdeum genes by hybridization-competition. Nucleic Acids Res. 9, 947-963.

10-Tien, M. and T.K. Kirk. 1983. Lignin-degrading enzyme from the hymenomycete Phanerochaete chrysosporium Burbs. Science 221, 661-663.

11-Tien, N. and T.K. Kirk. 1984. Lignin-degrading enzyme from Phanerochaete chrysosporium: purification, characterization and catalytic properties of a unique H_2O_2-requiring oxygenase. Proc. Natl. Acad. Sci. USA 81, 2280-2284.

12-Westermark, U. and K.-E. Eriksson. 1975. Purification and properties of cellobiose: quinone oxidoreductase from Sporotrichum pulverulentum. Acta Chem. Scand. B29, 419.

GENETIC ENGINEERING OF CELLULASE GENES FROM A FUNGUS

V.L.Seligy,M.J.Dove,M.Yaguchi,G.E.Willick and F.Moranelli
Molecular Genetics Section, Division of Biological Sciences,
National Resarch Council of Canada,K1A OR6

ABSTRACT- We have been studying the molecular genetic mechanisms
involved in the production of extra-cellular cellulases from the
white-rot,basidiomycetes fungus,Schizophyllum commune inorder
to better understand the enzymes and processes involved in the
biodegradation of cellulose to glucose. A report of recent
progress is made here   on aspects of the   structure of the
cellulase   enzymes,their   genes and their biosynthesis   and
excretion-related   processing . These studies have   direct
importance to global projects concering the development of large
scale production of cellulases and degradation of lignocellulosic
substrates for conversion to alcohol.

INTRODUCTION

There are several problems associated with development of an
efficient cellulose fermentation process. Stressing only the
biological side, the global trends in current research are to
obtain   a better characterization and production of the enzymes
involved in cellulose hydrolysis , to understand the regulatory
pathways involved in polysaccharide catabolism ,to select   new
or improved microbial strains with wider substrate adaptability
and to develop novel ,genetically engineered , organisms and
enzymes.

For the past three years we have carried out Industrial and
Bioenergy R&D related work with the general objective to use
recombinant DNA procedures for the purposes : 1) to isolate and
to characterize fungal genes coding for cellulolytic enzymes ;
2) to elucidate the secretory and regulatory mechanisms involved
in cellulase production; 3) to construct enzymes with novel
properties (e.g., increased or altered substrate specificity,
tolerance to physico-chemical changes of substrate environment,
introduction of bifunctional capability) and 4) to develop a
modified organisms (yeast or filamentous fungi) that might be
used to produce ethanol from residual oligosaccharides of paper
mill wastes (4-5).

MATERIALS AND METHODS

Our research is mainly concerned with the genetics of
cellulase   production   by   the   white-rot   basidiomycete
,Schizophyllum commune, one of the two organisms originally
chosen to be worked on in our section at NRCC (ref.1,2). The
major tasks ,techniques, strains and earlier progress that we
made on this organism have been described in detail(see ref.1-
7). Summarized here,the tasks centred mainly around determining

the type and performance of the enzymes required for the efficient degradation of the SSL oligosaccharides (1,2,4,6), the molecular characterization of the mechanism for hydrolysis used by the endoglucanase and exoglucanse (5-7) ,the ellucidation of regulatory mechanisms used by the cellulolytic organisms for temporal expression and extra-cellular production of the enzymes (1-9) ,the isolation of the genes coding for the enzyme CMC-1 and beta-glucosidase (4,7) , and the development of the necessary processes required for stable expression of the genes in an appropriate recipient yeast host (2,3).

RESULTS

a) Enzyme Structure and Specificity

The exact structure of the catalytic site for hydrolysis of the glucosidic bonds is unknown. The cellulases of Phanerochaete chrysosporium,Trichoderma reesei,Trichoderma koningii and S. commune (8,10,11) are particularly well studied. The attack on cellulose has been shown to involve the cooperative action of three types of 1,4-b-D-glucanases,including endo-1,4-b-D-glucanases(EC 3.2.1.4),exo-1,4-b-D-glucanases(EC 3.2.1.91),and 1,4-b-D-glucosidases(EC 3.2.1.21). The catalytic sites for the endoglucanases of S.commune (CMC-I and CMC-II) have been recently deduced by Dr.Yaguchi and associates from the sequence similarity with lysozyme(6,8) .Further similarities have been noted between CMC-I and II and T. reesei CBH I and EG1 (8). A general model for the catalytic site for both "alpha" and "beta" glucosidic bonds is plausible since the same distribution of amino acids has been recently noted for taka- amylase (12). Our recent success with the cloning and expression of a partially deleated ,but active,CMC-1 gene which codes mainly for the N-terminal region of CMC-1 ,containing the presumptive active site,should allow us to carry out production of a novel ,genetically engineered enzyme (vide infra).

b) Multiplicity of Enzymes

(i)Enzyme purification - Several extracellular proteins of S.commune have now been purified from submerged cultures containing cellulose(Solka Floc).These include two forms of xylanase ,two b-glucosidases(bG-I and bG-II),two Avicelases(Av-I and Av-II),two carboxymethylases(CMC-I and CMC-II) and two proteins with as yet unknown enzymatic activities(PX-I and PX-II). The physico-chemical properties of these enzymes has recently been described (9).

Preparations of these enzymes have been used for amino acid sequencing and to generate specific antibodies. We recently found that Beta-glucosidase can be purified in high-yield and efficiency using DEAE Biogel/HPLC chromatography. The enzymatic acivity can be resolved into two components, but the Mr of each is still heterogeneous. The enzyme specificity is to be tested against soluble oligosaccharides (2 to 6 mer). Lys-C protease digestion suggests all species may have at least some common sequence. The proteins are to be partially sequenced to assist in identification of the presumed glucosidase genes recently cloned by us.

(ii)Protein Sequence /Secretion/Modification- The enzymes CMC-I and CMC-II have been shown to be related by 16 additional amino acids at the N-terminus in CMC-I(6,8).This may be analogous to the late secretion of a glucoamylase by **Aspergillus niger**(13) which results from proteolytic cleavage at the N-terminus of the early form. Our recent studies on the biosynthesis and secretion of the bG,Av and CMC enzymes using specific antibody as probe are consistent with the idea three genes account for the secreted glucanases. The other species result from different glycosylation or proteolytic cleavage processing,which may occur during or after secretion (9).

We recently found the enzymes can be labelled with tritiated mannose. Each enzyme can be fractionated on concanavalin A-agarose column. Presence of tunicamycin results in the production of enzyme, nearly of similar size to the mRNA directed in vitro translation product. The non-glycosylated species is secreted. Our data suggests that part of the enzyme heterogeneity arises at the transcription/translation level and part during glycosylation and associated proteolytic processing (manuscript in preparation). More detailed studies on the mRNA of cellulases by in vitro translation indicate that the mRNA templates two species,the higher molecular weight component is always the most predominant.

c)Expression and Regulation

It is generally thought that the production of cellobiose,by the attack by enzymes secreted at a low level in the absence of glucose,is responsible for the induction of cellulases when the organism is presented with cellulose as the sole carbon source.It is also thought that the glucose produced from the attack of cellulases on the cellulose eventually acts to repress synthesis of the cellulases(10).

We previously reported that cellobiose was a potent inducer of the endoglucanase CMC-I(2). Using specific antibodies directed to bG,Av-I and CMC-I antigens as immunoprecipitation probes to monitor the relative levels of intracellular and extracellular cellulases of S.**commune** we recently showed that the production of cellulases is greatest when cellulose(Solka Floc) is used (9). Production drops about 10fold when the media contains carboxymethyl-cellulose and about 50-100fold when glucose is used .Data further obtained from immunoprecipitation experiments using radiolabelled proteins made in a rabbit reticulocyte cell-free translation system from messenger RNA (the poly A+ fraction)of mycelia grown on the same carbon sources indicate that the relative increase in secreted cellulase is directly related to an increase in the level of mRNA glucanase transcripts.The regulation of cellulase expression is therefore associated with transcription.

In the same publication we show that when the three cellulolytic activities are expressed in response to cellulose(Solkafloc), each activity is associated with at least two different Mr species. The members in each set have similar Mr(i.e., glucosidase= 100,000;Av-II=65,000; CMC=42,000).The endoglucanase and avicelase appear to be co-secreted. The glucosidase secretion can follow different induction curve which indicates that it may not be under the same regulation. The initiation of transcription of CMC-I and Av-II genes appears to be further co-ordinated as apparently is the secretion of the gene products (manuscript in preparation). This indicates that regulation acts at the level of gene transcription or of mRNA stability. Stability of mRNA coding for extracellular yeast invertase is linked in some way to one of the levels of expression involving translation, post-translational modification or secretion(14).

Preliminary investigations into the effects of varying nitrogen base concentration on cellulase production have been carried out. These studies indicate significant differences in the levels of total mRNA and specific mRNA production occur as the peptone concentration is varied in the presence of a constant amount of cellulose.For example,at levels less than 0.5% peptone ,cellular growth is severely curtailed ,the total mRNA level decreases dramatically and the relative levels of specific cellulase mRNAs appear to increase .It appears that nitrogen starvation or nutritional stress may be involved in triggering a shut-off of the growth-related transcription with little or no affect on the cellulase transcription.These conclusions are not too unlike those proposed for genetic regulation of secondary metabolism in bacteria (15). However, the relative concentrations of the specific mRNA will have to be measured inorder to confirm the differences seen in the levels of translation capacity.This can be done now by using the cloned genes as sequence detection probes.

d)Isolation of CMC-1 and Glucosidase Gene Sequences

(i)Gene Cloning methodology- Initially,like other groups working on fungal gene cloning(11), we used all of the gene screening methods,including synthetic oligonucleotides constructed from predicted codon usage based on the amino acid sequence of CMC-1. The result was the isolation of a few hundred gene sequences which were related to growth adaptaion on cellulose (2,3). By switching to a gene fusion expression vector,we isolated several clones carrying partial DNA sequences of CMC-I and b-glucosidase. The plaques harboring the clones produced nanogram quantities of fused protein made by the fused S.commune-E.coli gene sequences which could be detected by radioimmune assay using respective specific antibodies as probe.

(ii)Gene Isolation/Characterization/Expression- We have characterized these clones with respect to the fused protein product and DNA sequence . The sequencing of the glucosidase cDNA clone has been started. Sequencing of the CMC-1 is approx. two-thirds complete. The coding region has been located in relation to the amino acid sequence of CMC-1 (manuscript in preparation). We have cloned the corresponding genomic sequence of CMC-1. Our current work is directed to sequence mapping and attempts to sub-clone and express the cDNA and genomic genes in E.coli and yeast. Through a novel rearrangement of the genomic sequence of CMC-1 we have obtained enzymatic expression in E.coli.(manuscript in preparation). This modified gene sequence is now being studied in yeast.

DISCUSSION

The potential importance of cellulases for a variety of commercial uses,including alcohol production, has led to new approaches to develop cellulolytic industrial organisms. We have carried out an extensive study on the molecular genetics of cellulases of the white-rot fungus,S.commune. Our studies ,which are still on-going,indicate that a number of molecular changes in gene expression occur when S.commune is grown on cellulose rather than glucose. The number and kind of cellulases so far identified are similar to those produced by other cellulolytic fungi that have been studied (10,11).

The studies on protein structure and function using amino acid sequencing and specific antibodies as probes have been useful in revealing the likely composition of the catalytic site used in hydrolysis of the 1,4 glucosidic bonds and the possible origins of enzyme variants. Combining this information with other studies (12) a general model for alpha and beta 1,4 hydrolysis may be postulated.

The biosynthetic pathway and mechanisms involved in extra-cellular production of cellulases appear to parallel the same events observed for extra-cellular proteins of yeast and other filamentous fungi (see 3,9,13-14).Some aspects of cellulase gene regulation have been uncovered through use of immuprecipitation and in vitro translation of specific mRNA. We expect to improve on these details by use of the cloned genes as hybridization probes.

The most rapid way to locate a clone harboring a specific gene is through detection of its product by genetic complementation or enzymatic assay ,if the gene is functional, or by radioimmune assay,if the gene is only partially functional. These approaches have been used to obtain cellulase genes from bacteria (15-16) and now from S.commune. The screening methods are generally much easier to implement than those required to detect the gene(DNA) sequence,which involves the technique of nucleic acid hybridization and tedious construction of radioactive DNA or RNA sequences as probes. These nucleic acid probes are usually at best only enrichments of gene sequences,and require additional techniques to identify the gene by its product or DNA sequence.

It is still too early to comment on how the genes are induced to synthesize cellulases or what the genetic organization and linkage of the cloned CMC-1 and glucosidase genes might be.

Preliminary comparisons indicate that there is very little sequence homology of fungal cellulases outside of the few residues defining the active site. Unlike bacteria, the coding sequences of fungal genes are disrupted by noncoding regions.This may be the reason why the cloned CBH1 gene of Trichoderma was not expressed in E.coli (11). However,through major deletions of the genomic CMC-1 gene we have obtained enzymatic expression in bacteria. It is not clear whether split-genes from one fungal species can be processed by another. The glucoamylase of both Aspergillus amouri and niger are nearly identical in structure.The enzyme could only be expressed by fusing just the coding sequence (cDNA) with yeast transcription signals(18a). However, a full genomic sequence of glucosidase from Aspergillus niger apparently does express in yeast at low levels (17). This lab also claimed a similar success for expression of CBH1 genomic sequence (18c) which is slightly contradictory to the approach taken by the Cetus group(18b).Our present work is directed to the expression of both the full genomic and the truncated CMC-1 gene in yeast. A second option will be to fuse the cDNA (mRNA) copies of CMC-1 and glucosidase with yeast transcription signals in a similar manner tried with moderate success using the CBH1 and EG1 cDNA(18b).

ACKNOWLEDGEMENTS

We thank J.R.Barbier,C.Roy and F.Rollin for technical help.We further extend our appreciation for support from NRCC , Energy Division and appreciation for the interactions with colleagues at PAPRICAN,M.Desrochers,L.Jurasek, M.Paice .

REFERENCES

1. James,A.P.Seligy,V.L.Thomas,D.Y.Yaguchi,M.Willick,G.E.Morosoli,M. Desrocher,M. and Jurasek,L. Proc.Third Bioenergy R&D Seminar,Ottawa,1981,135-139.

2. James,A.P.,Seligy,V.L.,Thomas,D.Y.,Yaguchi,M.,Willick,G.E., Morosoli,R.,Crosby,W.,Desrochers,M.and Jurasek,L. Proc. Fourth Bioenergy R&D Seminar,Winnipeg,1982,425-429.

3. Seligy,V.L.,Barbier,J.R.,Dimock,K.D.,Dove,M.J.,Moranelli,F., Morosoli,R.,Willick,G.E. and Yaguchi,M. In,Gene Expression in yeast. Proc. ALKO Yeast Symposium,Foundation for Biotechnical and Industrial Fermentation Research 1(1983)167-185

4. Willick,G.E.,Seligy,V.L.,Veliky,I.and Jurasek,L.TAPPI R&D Conf.Proc.,1982,307-315

5. Paice,M.,Jurasek,L. and Desrochers,M.TAPPI R&D Conf.Proc., 1982,103-106

6. Yaguchi,M.,Roy,C.,Rollin,F.,Paice,M. and Jurasek,L..Biochem. Biophys.Res.Commun.116(1983)408-411.

7. V.L.Seligy,M.J.Dove,C.Roy,M.Yaguchi,G.E.Willick,J.R.Barbier, L.Huang and F.Moranelli. In, Proc.Fifth Bioenergy R&D Seminar, Ottawa,1984. (in press)

8. Paice,M.,Desrochers,M.,Rho,D.,Jurasek,L.,Roy,C.,Rollin,F.
    DeMiguel,E. and Yaguchi,M.Bio/technology 2(1984)535-539

9. Willick,G.E.,Morosoli,R.,Seligy,V.L.,Yaguchi,M.,Desrochers,M.
    J.Bact.159(1984)294-299

10. Eriksson,K.-E.Cellulases of Fungi.Trends in Biology of
    Fermentations.Plenum Press,1981.pp19-31.

11. Shoemaker,S.,Schweickart,V.,Ladner,M.,Gelfand,D.,Kwok,S,
    Myambo,K.and Innis,M.Bio?technology,1(1983)691-696.

12. Matsuura,Y.,Kusunoki,M,Harada,W,and Kakudo,M.J.Biochem.
    95(1984)697-702.

13. Boel,E.,Hjort,I.,Svensson,B.,Norris,F.,Norris,K.E.and
    Fiil,N.P. The EMBO J.3(1984)1097-1102.

14. Schekman,R.and   Novick,P.In,Molecular   Biology   of
    Saccharomyces,CSH Press,1982,pp361-398.

15. Gilkes,N.,Langsford,M.L.,Kilburn,D.G.,Miller,R.C.   and
    Warren,A.J.J.Biol.Chem.259(1984) 10455-10459

16. Beguin,P.,Cornet,P. and Millet,J. Biochimie 65(1983)495-500

17. Penttila,M.,Nevalainen,H.,Raynal,A.and Knowles,J. Mol.Gen.
    Genet.194(1984)494-499

18. All papers presented at 12th International Conference on
    Yeast Genetics and Molecular Biology,Edinburgh,September,1984.
    a)Innis,M.A. et al Abstract G11,
    b)Glefand,D.et al,Abstract G14,
    c)Nevalainen,M. et al, Abstract G15

The development of Azotobacter as a bacterial fertilizer by the introduction of exogenous cellulase genes

Bernard R Glick, J.J. Pasternak and Heather E. Brooks
Department of Biology
University of Waterloo
Waterloo, Ontario
Canada N2L 3G1

Aerobic, nitrogen fixing, free-living soil microorganisms such as Azotobacter may, with the aid of genetic manipulation, provide an additional source of renewable nitrogen for enhancing crop yields. Azotobacter has a number of biological attributes to perform the task of an effective bacterial fertilizer. It is found in temperate zones, it is hardy and not soil-specific and its outer 'gummy' coat should enable it to be used as a seed dressing. However, its nitrogenase (i.e, the enzyme that fixes nitrogen) is highly susceptible to oxygen. This inhibition can be overcome if the organism attains high levels of respiration which 'burns off' excess internal oxygen. In order to maintain a high level of respiration, we plan to determine if the endoglucanase and cellobiase genes of other gram negative microorganisms, once cloned into Azotobacter will provide large amounts of metabolizable carbon (glucose) and whether the levels of both respiration and nitrogen fixation are elevated. Here, we discuss in detail the rationale behind the strategy for genetically engineering a cellulolytic Azotobacter ; outline plans to create a model system to test the basic concept underlying the development of a cellulolytic Azotobacter and describe a protocol for obtaining high frequency transformation of Azotobacter with the broad host range plasmids pRK2501, RSF1010 and pGSS15.

Annually, more than 100 million tons of nitrogen are needed to sustain global food production. About half of this nitrogen is derived from synthetic fertilizers. The remainder is obtained, for the most part, from diazotrophic (i.e., nitrogen fixing) microorganisms such as Rhizobia, Frankia, Azospirillum, Azotobacter and cyanobacteria. Certainly, as the world population increases, the requirement for fixed nitrogen for crop production will increase concomitantly (1,2).

As long as petroleum was in ready supply and chemical fertilizers were cheap there was little incentive to develop alternate sources of fixed nitrogen. However, while there has been a brief respite from escalating oil prices, it has become apparent that there are advantages to developing sources of fixed nitrogen that do not depend so heavily on a diminishing and non-renewable energy supply. With this in mind, we have set out to develop Azotobacter, a free-living nitrogen fixing microorganism, as a bacterial fertilizer.

The purposeful use of a diazotrophic microorganism, such as Azotobacter, for enhancing crop growth (e.g. as either a soil or seed inoculant), has been somewhat limited. The mass of data collected by Soviet scientists in the 1950's and 1960's on the use and effects of bacterial fertilizers has generally been dismissed because the experimental methodology lacked appropriate controls and input mixtures were poorly defined. While some of the reported results were intriguing, these early experiments did not provide a clear vision of how to proceed. Nevertheless, well-designed experiments have since demonstrated that various bacterial species, including Azotobacter, are effective in increasing crop yields both in laboratory and field trials (3-6). A variety of strategies have been proposed to increase the efficiency of Azotobacter as a bacterial fertilizer, including the development of mutants which fix excessive amounts of nitrogen (4).

Azotobacter has some intrinsic biological properties that could be utilized in the development of an effective bacterial fertilizer. It is found in temperate zones, its sticky outer coat enables it to be used as a seed dressing and it is quite hardy and not soil specific. However, although it can grow in the presence of oxygen, its nitrogenase, the enzyme responsible for the fixing of nitrogen, is irreversibly inactivated by oxygen. To protect its nitrogenase from this toxic effect, Azotobacter maintains a very high rate of respiration, in effect 'burning off' excess cellular oxygen. This unique mechanism of protection entails adaptation of the entire respiratory system to the varying levels of oxygen in the soil (7). Although the mechanism of this 'respiratory protection' is complex and many of the details are not well understood, a schematic representation of the overall phenomenon is presented in Fig. 1. In this depiction, the level of available oxygen within the bacteria (i.e., cellular oxygen) is such that nitrogenase activity is dramatically curtailed. However, if the organism increases its respiratory rate due to an increased availablity of glucose, cellular oxygen would be used and the internal supply would therefore be diminished; consequently, relieving the inhibition of nitrogenase activity. Moreover, in its natural setting, the extent to which Azotobacter can colonize the rhizosphere is limited because of the large amounts of metabolizable carbon that it requires. For this reason Azotobacter does not compete efficiently with other soil microorganisms which also use plant root exudates as a major carbon source.

We have hypothesized that the ultimate efficiency and usefulness of Azotobacter as a bacterial fertilizer depends upon providing it with a sufficient amount of metabolizable carbon so that it would be (i) competitive with other microorganisms in the rhizosphere and (ii) would not be thwarted in its fixation of nitrogen by the presence of cellular oxygen. This concept is illustrated schematically in Fig. 2. Unlike the conditions shown in Fig. 1, a cellulolytic Azotobacter would be able to produce large amounts of metabolizable carbon (glucose) in its immediate vicinity from soil cellulosic material thereby increasing the internal level of glucose. In turn, a considerable amount of cellular oxygen would be utilized in respiration. Consequently, the inhibition of nitrogenase by oxygen would be relieved and the organism would fix more nitrogen as well as being more competitive with other soil microorganisms. Such a genetically engineered Azotobacter, theoretically, might be an effective and inexpensive bacterial fertilizer.

It has been previously noted that in the presence of cellulose-degrading microorganisms, the concentration of Azotobacter in the rhizosphere is significantly increased (8). Moreover, cellulose-degrading microorganisms and Azotobacter act synergistically to stimulate plant growth (9). The natural occurrence of both cellulolytic activity and nitrogen fixation in a single organism has until recently not been observed. However, a novel cellulolytic nitrogen fixing microorganism has recently been found in a gland of marine wood-boring shipworms (10). This discovery suggests that there is no intrinsic biological barrier to the development of an organism by genetic manipulation that is capable of both degrading cellulose and fixing nitrogen.

Cellulose is biologically degraded by a set of enzymes that is commonly referred to as 'cellulase'. The cellulase complex may consist of as many as four different enzymatic activities including exoglucanase, cellobiohydrolase, endoglucanase and β-glucosidase (Fig. 3). Much of the available evidence suggests that different microorganisms use various combinations of these enzymes to hydrolyze cellulose. What is particularly germane from our point of view is that there are several cellulolytic bacteria that are able to hydrolyze cellulose to glucose using only an endoglucanase and a β-glucosidase. Therefore, the initial targets in our

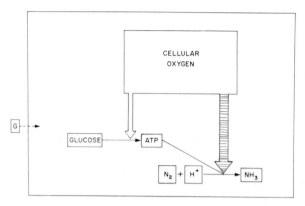

AZOTOBACTER

Fig. 1. Schematic representation showing how the level of cellular oxygen, if
large, depresses nitrogen fixation in normal Azotobacter. The amount of
metabolizable carbon, usually glucose (G), limits the amount of cellular
oxygen that is utilized for respiration. Thus, the lower the amount of
available metabolizable carbon, the lower the extent of respiration, the
higher the concentration of oxygen that will be present in the cell and
the more likely that the nitrogenase will be inhibited.

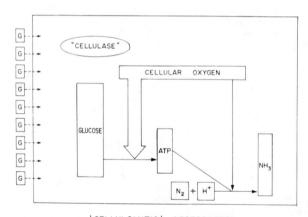

'CELLULOLYTIC' AZOTOBACTER

Fig. 2. Schematic representation showing the expected consequences of genetically
engineering a 'cellulolytic' Azotobacter. In contrast to Fig. 1, the
amount of available metabolizable carbon is relatively high due to
cellulolytic activity so that the level of respiration is also
relatively high while the concentration of cellular oxygen is lowered.
Under these conditions, nitrogenase activity is not readily inhibited by
cellular oxygen and the extent of nitrogen fixation is enhanced.

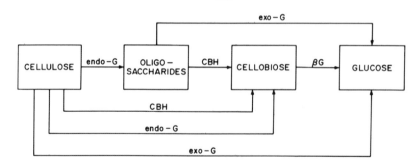

Fig. 3. Schematic representation of the enzymatic hydrolysis of cellulose. exo-G, exo-β-1,4-glucanase; endo-G, endo-β-1,4- glucanase; CBH, exo-β-1,4-glucanase (cellobiohydrolase); G, β-glucosidase (cellobiase)

efforts to isolate genes coding for proteins which degrade cellulose and which can be transformed into Azotobacter are those bacterial genes that code for endoglucanase and a β-glucosidase.

We have recently constructed clone banks of some gram negative aerobic cellulolytic soil microorganisms into the E. coli plasmid, pBR322. The clone banks were made by ligating Sau3A-digested genomic DNA to BamHI-digested pBR322. These clone banks are being screened for the expression of endoglucanase activity using the Congo Red plate assay devised by Teather and Wood (11) as well as by DNA hybridization using a cloned Bacteriodes succinogenes endoglucanase gene (kindly provided by Bill Crosby and Dave Thomas of the National Research Council of Canada). Preliminary data indicate that the Bacteriodes succinogenes gene hybridizes to genomic DNA from a cellulolytic pseudomonad under relatively non-stringent conditions. Further work will establish whether this is, indeed, an endoglucanase gene. In addition, the clone banks will be screened for the expression of β-glucosidase activity using MacConkey-cellobiose indicator plates and by DNA hybridization using a cloned β-glucosidase gene from Escherichia adecarboxylata (kindly supplied by Richard Armentrout of Syntro Corporation). In order to test the feasibility of creating a free-living cellulolytic diazotroph, we are in the process of subcloning the endoglucanase gene from B. succinogenes and the β-glucosidase gene from E. adecarboxylata into broad host-range plasmid vectors which, in turn, can be introduced into Azotobacter. Such an organism can be used as a model system for optimizing the development of Azotobacter as a bacterial fertilizer and for testing the the concept in general.

In order to manipulate an organism genetically it is important to be able to readily introduce foreign genes to alter and/or augment its biological capabilities. The process of genetic transformation should be efficient and to produce permanent genetic changes the exogenous genetic material must be stably maintained by the recipient organism. We have, therefore, developed a protocol for the transformation of Azotobacter vinelandii in liquid media with plasmid DNA. The transformation of Azotobacter was achieved with three different broad host-range plasmids (i.e. pRK2501, RSF1010 and pGSS15). Briefly, plasmid pRK2501 is 11.1 kb, carries genes coding for resistance to the antibiotics tetracycline and kanamycin, is a member of the P1 incompatibility group and is a derivative of the naturally-occurring 56.4 kb plasmid RK2. Plasmid RSF1010 is a naturally-occuring 8.5 kb plasmid that carries genes that code for resistance to the antibiotics sulfonamide and streptomycin and is a member of incompatibility group Q. Plasmid pGSS15 is an 11.3 kb derivative of

RSF1010 and carries genes which code for resistance to ampicillin and tetracycline. The recipient microorganism was a laboratory strain of <u>Azotobacter vinelandii</u> (ATCC No. 12837), which as far as we are able to ascertain, does not possess endogenous plasmids. The transformation procedure entailed growing <u>Azotobacter</u> in a transformation medium (1.9718 g $MgSO_4$, 0.0136 g $CaSO_4$, 1.1 g $CH_3CO_2NH_4$, 10.0 g glucose, 0.25 g KH_2PO_4 and 0.55 g K_2HPO_4 per litre) which is similar to the one developed by Page and von Tigerstrom (12, 13). Aliquots of <u>Azotobacter</u> were removed at various times after the initiation of cell growth and approximately 1.6×10^8 cells were suspended in 50 μl of fresh transformation media. Routinely, 50 μl of plasmid DNA (22 μg/ml) was then added followed by the addition of 300 μl of fresh transformation medium. After 30 min at 30°C, the suspension was centrifuged (12,000 x g), the pellet was resuspended in 400 μl of fresh transformation medium and incubated at 30°C for 60 min. Aliquots of this cell suspension were plated onto agar plates of <u>Azotobacter</u> growth medium (0.2 g $MgSO_4$, 0.1 g $CaSO_4$, 0.2 g KH_2PO_4, 0.8 g K_2HPO_4, trace amounts of $FeCl_3$ and $NaMoO_4$, 20.0 g sucrose, and 0.5 g yeast extract per litre) to which the selective antibiotic either had or had not been added (Fig. 4).

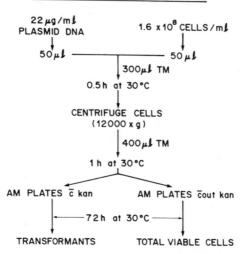

Fig. 4. Summary of the protocol that was devised for the transformation of <u>Azotobacter vinelandii</u> with plasmid DNA. TM, transformation medium; AM; growth medium; kan, kanamycin. In this example, kanamycin was used to select transformants that carried pRK2501.

This transformation protocol is highly effective and transformation frequencies (i.e., number of transformants/μg pRK2501 DNA/viable cell) of approximately 10^{-2} were attained. Other workers have noted that <u>Azotobacter</u> is capable of taking up chromosomal DNA only during a narrow time span in the cell growth cycle (13-15). To test whether the uptake of plasmid DNA is also confined to a specific phase of the cell growth cycle, cells were transformed at different stages of the growth cycle. In transformation medium, growth is biphasic with a sharp break occurring when the absorbance value (OD620) of the culture is 0.1-0.2. This abrupt change in the growth curve corresponds to the release of a fluorescent green pigment (presumably an Azotobacter siderophore (16)) into the medium. For each of the three plasmids that were studied, transformation was found to be independent of the stage of cell growth. Plasmid pRK2501 gave a uniform transformation frequency (1-5 x 10^{-2}) throughout the cell growth cycle; whereas, pGSS15 and RSF1010 showed some scatter at or below absorbance values corresponding to the sharp break in the growth curve. The transformation frequency for pGSS15 was approximately 2 x 10^{-2} while for RSF1010 it ranged from 3 x 10^{-4} to 4 x 10^{-2} Thus, on the basis of these experiments, it is inferred that Azotobacter is constitutive for plasmid DNA transformation.

When Azotobacter was transformed with increasing amounts of pRK2501 DNA, the transformation frequency was proportional to the amount of added plasmid DNA (Fig. 5). Thus, on the one hand, transformation frequency can be quite high (e.g., >30% of the cell population becomes transformed) while, on the other, the efficiency of uptake of plasmid DNA is very low (as evidenced by the levels of plasmid DNA that

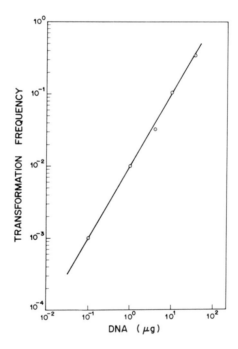

Fig. 5. The effect of the concentration of pRK2501 DNA on the transformation frequency of Azotobacter vinelandii. Cells were grown to an absorbance at 620 nm of 0.4 in transformation medium and transformed following the standard protocol. Transformation frequency is the number of transformants per μg pRK2501 per total number of viable cells. The correlation coefficient for the best-fit line is + 0.998.

are required to attain high transformation frequencies). The mechanism by which Azotobacter sequesters exogenous DNA and transports it to the interior of the cell is not known. Based on studies of Azotobacter transformation using chromosomal DNA, it has been postulated (17) that there may be two different types of receptor sites for DNA on the cell surface. One site binds DNA throughout the cell growth cycle and a second site not only binds DNA but also facilitates its transport into the cell. Activation of the latter site occurs when the cells are competent to take up exogenous DNA. Our studies are consistent with the possibility that plasmid DNA binds, albeit inefficiently, to a receptor site and once bound is then taken up by the cell in the absence of any competence-inducing factor(s).

When Azotobacter that had been transformed with plasmid DNA were plated onto growth medium, without the addition of a selective antibiotic, and incubated at 30°C overnight, two distinct colony morphologies were observed (Fig. 6). The larger colonies arise from transformed cells while the smaller colonies are from non-transformed cells. This difference was ascertained by replica plating of disparate types of colonies onto antibiotic-containing medium. After growth of both transformed and non-transformed cells on this medium for approximately 7 days, non-transformed and transformed cells attain similar colony sizes although the colony morphologies of transformed and non-transformed cells are still visibly different. In the latter case, the transformed cells seem to have increased amounts of capsular slime. When re-tested, after many rounds of subculturing on fresh medium, the difference in colony morphology is maintained. In liquid culture, the transformed cells grow more rapidly initially but level off at a lower absorbance value than the non-transformed cells (Fig. 7). In sum, the presence of plasmid DNA per se alters the normal physiology of the cells, probably by creating an increased metabolic load. However, in spite of this 'load', we have recently been able to show, using the acetylene reduction assay (18), that transformed cells retain their ability to fix nitrogen.

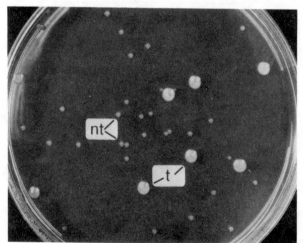

Fig. 6. Colony morphology of transformed and nontransformed cells when grown on Azotobacter growth medium for 1 day at 30°C. Large colonies (t) comprise cells transformed by pRK2501 and are resistant to kanamycin. Nontransformed cells (nt), when tested, did not grow on media that was supplemented with kanamycin.

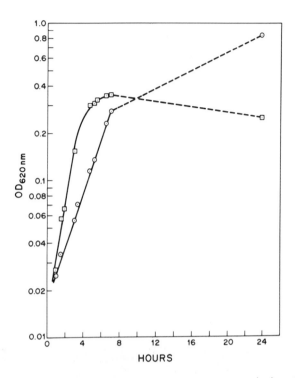

Fig. 7. Representative growth curves of plasmid-transformed (□) and nontransformed
(O) <u>Azotobacter</u> <u>vinelandii</u> in growth medium (AM).

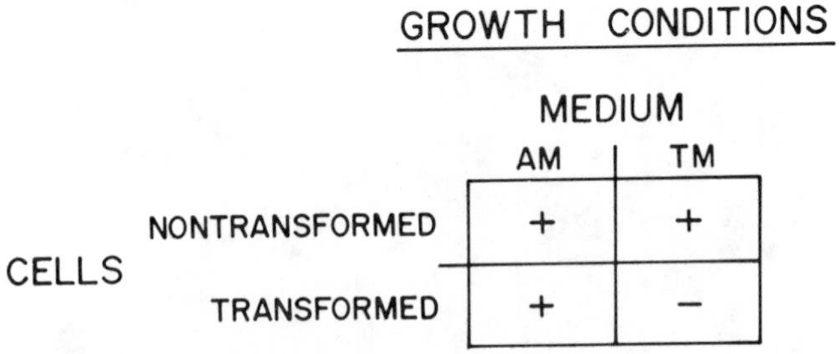

GROWTH CONDITIONS

MEDIUM

	AM	TM
NONTRANSFORMED	+	+
TRANSFORMED	+	−

CELLS

Fig. 8. Rationale for the direct testing of plasmid retention by transformed
<u>Azotobacter</u> in the absence of selecting agents. Transformed cells do
not grow on transformation medium (TM); whereas, both transformed and
nontransformed cells grow on growth medium (AM) and nontransformed cells
grow on TM medium. Transformed cells are grown on AM medium for a
number of generations and then plated onto TM agar. The formation of
any colonies on TM medium signifies that the plasmid was not retained
during growth in growth medium. In a number of trials, plasmid
retention was 100%.

Because transformed <u>Azotobacter</u> does not grow on the transformation medium, a simple test to determine whether the plasmid is retained in absence of selection pressure is feasible. The rationale for this test is presented schematically in Fig. 8, where AM denotes the growth medium and TM the transformation medium. Briefly, transformants are grown for a number of generations in liquid AM media in the absence of antibiotic. At various times aliquots are withdrawn and plated onto TM media. Any colony formation on TM media would signify loss of the plasmid; that is, a reversion of the transformed state. In our experience, the plasmid is retained by all of the cells in the absence of any selective pressure for more than ten cell generations. In addition, since we are able to recover plasmid DNA from transformed cells, the stable transformed phenotype does not result from the integration of the plasmid DNA into the <u>Azotobacter</u> chromosome.

The ultimate aim of this work is to test the efficacy of a genetically engineered cellulolytic <u>Azotobacter</u> as a bacterial fertilizer by enhancing the ability of such an organism to provide fixed nitrogen to growing plants. However, it is questionable whether free-living nitrogen-fixing, soil microorganisms stimulate plant growth solely by supplying fixed nitrogen (3, 19, 20). It is likely that the relationship between plants and soil microorganisms is more complicated. Thus, these kinds of soil microorganisms may stimulate plant growth by providing growth regulatory substances, especially during early stages of plant development, as well as fixed nitrogen, even in small amounts (6). Regardless of the mechanism(s) by which diazotrophic microorganisms stimulate plant growth, it is certainly worthwhile to genetically manipulate these organisms since in their presence plant growth is clearly enhanced. Such genetic strategies are necessary if effective biological alternatives to energy intensive fertilizers are to be developed.

References

1. Postgate, J.R. Phil. Trans. R. Soc. Lond. B 290: 421-425, 1980.
2. Worne, H. Genetic Eng. News 3: 6-7, 1983.
3. Brown, M.E. In Bacteria and Plants, eds. Rhodes-Roberts, M.E. and Skinner, F.A., pp. 25-41, Academic Press, 1982.
4. Gordon, J.K. and Jacobson, M.R. Can. J. Microbiol. 29: 973-978, 1983.
5. Baldani, V.L.D., Baldani, J.I. and Dobereiner, J. Can. J. Microbiol. 29: 924-929, 1983.
6. Kapulnik, Y., Sarig, S, Nur, I. and Okon, Y. Can. J. Microbiol. 29: 895-889, 1983.
7. Robson, R.L. and Postgate, J.R. Ann. Rev. Micriobiol. 34: 183-207, 1980.
8. Jensen, H.L. Proc. Linn. Soc. N.S.W. 65: 1-122, 1940.
9. Mishustin, E.N. and Shilnikova, V.K. Soil Biology. pp. 72-109, UNESCO, 1969.
10. Waterbury, J.B., Callaway, C.B. and Turner, R.D. Science 221: 1401-1403, 1983.
11. Teather, R.M. and Wood, P.J. Appl. Environ. Microbiol. 43: 777-780, 1982.
12. Page, W.J. and von Tigerstrom, M. J. Bacteriol. 139: 1058-1061, 1979.
13. Page, W.J. and von Tigerstrom, M. Can. J. Microbiol. 24: 1590-1594, 1978.
14. Page, W.J. Can. J. Microbiol. 28: 389-397, 1982.
15. Reusch, R. and Sadoff, H.L. J. Bacteriol. 156: 778-788, 1983.
16. Page, W.J. and Huyer, M. J. Bacteriol. 158: 496-502, 1984.
17. Doran, J.L. and Page, W.J. J. Bacteriol. 155: 159-168, 1983.

134

18. Hardy, R.W.F., Holsten, R.D., Jackson, E.K. and Burns, R.C. Plant Physiol. 4
    1185-1207, 1968.
19. Brown, M.E. and Carr, G.R. J. Appl. Bacteriol. 56: 429-437, 1984.
20. Okon, Y., Heytler, P.G. and Hardy, R.W.F. Appl. Environ. Microbiol. 46
    694-697, 1983.

Acknowledgement

The original research cited in this paper was supported by the Natural
Sciences and Engineeing Research Council of Canada.

Butanediol Production from Cellulose and Hemicellulose

Ernest K.C. Yu and John N. Saddler
Biotechnology Department, Forintek Canada Corporation
800 Montreal Road, Ottawa, Ontario, Canada K1G 3Z5

ABSTRACT

The bioconversion of cellulose and hemicellulose components of lignocellulosic substrates to butanediol was investigated. All the major sugars present in the cellulose and hemicellulose hydrolyzates were efficiently utilized by <u>Klebsiella</u> <u>pneumoniae</u> to produce 2,3-butanediol. The fermentation process was studied in a chemically defined medium under various culture conditions and optimized for maximal final butanediol concentrations (g/L) and bioconversion yields (g of product per g of sugar used). By using a fed-batch approach final butanediol levels exceeding the concentrations considered desirable for economic product recovery were obtained from commercial sugars at near theoretical yields. Butanediol was also successfully produced from wood and agricultural residues after acid or enzymatic hydrolysis. A combined enzymatic hydrolysis and fermentation approach was adopted to enhance the efficiencies of the enzymatic hydrolysis of hemicellulose and cellulose carbohydrates. The approach was optimized with aspenwood xylan and solka floc and demonstrated technical feasibility for the direct utilization of steam-exploded substrates (i.e., the combined utilization of hemi-cellulose and cellulose components of different substrates).

INTRODUCTION

The production of 2,3-butanediol from lignocellulosic substrates has recently been considered a potentially viable approach for the bioconversion of biomass residues to fuels and chemicals (Flickinger, 1980; Jansen <u>et al</u>, 1984; Rosenberg, 1980). Various microorganisms are known to produce butanediol as a major fermentation end-product. To date, one of the most promising candidate appears to be the bacterium <u>Klebsiella</u> <u>pneumoniae</u> (previously known as <u>Aerobacter aerogenes</u>). There are several advantages to butanediol fermentation by this organism which render the process attractive. One of the most important advantages is that the organism is nutritionally versatile and capable of utilizing all the major sugars known to be present in cellulose and hemicellulose hydrolyzates (Yu and Saddler, 1982a). Moreover, the organism carrying out the fermentation is a fast-growing, facultative anaerobe, and is consequently relatively easy to work with. It can produce butanediol at near theoretical efficiencies from both the pentose and hexose sugars. Furthermore, the relatively non-toxic nature of the diols enables the accumulation of high final product concentrations which is important for the economics of product recovery and ultimately the whole bioconversion process. The production of butanediol is also appealing due to the potential importance of the product as a fuel or chemical feedstock. The potential uses of butanediol are diverse (Long and Patrick, 1963; Ledingham and Neish, 1954). It can be readily dehydrated to methyl ethyl ketone as an industrial solvent, and to butadiene for the manufacturing of synthetic rubber. More recently, interest in butanediol production has emphasized its use as a feedstock for the production of polyesters and polyurethane resins (precursors of synthetics and plastics), consequently extending its use as a valuable specialty chemical. At present, the major problem facing the bioconversion process appears to be mainly in the economic competitiveness of the product. If the sugars required for the fermentation process could be obtained from inexpensive and abundant biomass residues, such as wood and agricultural wastes, the overall cost of the process could be substantially decreased. The economics of the process could also be further enhanced if both the cellulose and the hemicellulose components of the biomass substrates were efficiently converted to useful products. We are therefore interested in establishing a process for the complete utilization of lignocellulosic substrates to produce butanediol. This paper presents a brief review of the work carried out in our laboratory on the optimization of butanediol fermentation by K. pneumoniae and our attempts at establishing a potential process, including pretreatment by steam-explosion, enzymatic hydrolysis, and fermentation, for the production of butanediol.

MATERIALS AND METHODS

Microorganism and medium.

Klebsiella pneumoniae (formerly Aerobacter aerogenes NRRL B-199; K. oxytoca ATCC 8724) was obtained from National Research Council of Canada (NRCC 3006). Methods for maintenance of stock cultures and routine checks for strain purity were previously described (Yu and Saddler, 1982a). The organism was grown in a chemically defined medium and incubated under finite air conditions at 30°C with shaking (150 rpm) (Yu and Saddler, 1982b; 1983). Procedures involved in acclimatizing the organism to high initial substrate concentrations (15%, w/v) and the fed-batch approach were previously described (Yu and Saddler, 1983).

Enzyme preparations.

Crude cellulase and xylanase enzyme preparations (crude filtrates) were used in this study for the enzymatic hydrolysis of cellulose and hemicellulose carbohydrates. Trichoderma harzianum E58 from the Forintek Culture Collection was grown on 2% (w/v) solka floc in Vogel's medium in a 30-L New Brunswick Microferm Fermentor at 28°C for 6 days (Saddler, 1982). The resulting cultures were filtered through Whatman glass fiber filter and then concentrated by membrane filtration with an exclusion limit of 10,000 Daltons (Pellicon Casette System, Millipore Ltd., Mississauga, Ontario, Canada). The Pellicon retentates were adjusted to pH 6.5, sterilized by passage through a 0.22-0.45 μm Millipore membrane and then added to pre-sterilized media.

Substrates.

Xylan standards were prepared by alkali treatment of extractive-free aspen sawdust (Jones et al, 1961; Koshijima et al, 1965). Aspenwood chips which had been previously vacuum impregnated with 0.2% sulfuric acid and various agricultural residues (wheat and barley straw, corn stover) were exposed to saturated steam at 220°C or 240°C for varying durations (10s to 120s) prior to explosion in a previously described 2-L, high pressure digester (Saddler et al, 1982). The steam-exploded substrates were extracted with water (substrate to water, 1:10, w/w) at room temperature for 2 h. The water-extracts were collected and concentrated by rotary evaporation. The original steam-exploded samples (SE), the water-soluble extracts (WS) and the water-insoluble residues (WI) were then analyzed for their carbohydrate content and subsequently hydrolyzed and fermented.

Combined hydrolysis and fermentation (CHF) studies.

Procedures used in the CHF approach for the utilization of wood and agricultural residues were described previously (Yu et al, 1984a; 1984b). Addition of enzymes (Pellicon retentates) was carried out at the onset of the CHF process, and the desired final enzyme (xylanase or endoglucanase) levels were adjusted depending on the concentration of hemicellulose or cellulose used.

Analytical methods.

Pentosans were analyzed by the method outlined in TAPPI standard Y223-os-71. Acid-insoluble lignin was assayed by the method outlined in TAPPI T222-os-75. Hexosan was analyzed by the anthrone reaction (Shields and Burnett, 1970).

Soluble protein was determined by the method of Lowry et al (1951) using bovine serum albumin as standard. Filter paper activity was determined by the method of Mandels et al (1976). Xylanase activity was determined as previously described (Saddler et al, 1983). Endoglucanase (carboxymethylcellulase) and β-glucosidase (salicinase) activities were assayed under conditions previously established (Saddler et al, 1982a). Xylobiase (β-xylosidase) activity was determined using p-nitrophenyl xylopyranoside reagent (Dekker, 1983).

Culture supernatant fluids were analyzed for volatile fermentation products by gas chromatography under conditions previously defined (Yu and Saddler, 1982b). Reducing sugars were assayed with dinitrosalicylic acid reagent (Miller, 1959). Mono-saccharides were assayed by high pressure liquid chromatography (Wentz et al, 1982).

RESULTS AND DISCUSSION

Initial studies were carried out to establish the culture conditions for optimal butanediol production by Klebsiella pneumoniae. When grown in a chemically defined medium, the organism could readily utilize all the major sugars known to be present in biomass hydrolyzates (Yu and Saddler, 1982a). The bioconversions (particularly with D-xylose) were enhanced in the presence of an organic buffer, MES (2-[N-morpholino]-ethanesulfonic acid, pKa 6.1) and under finite air conditions (Table 1).

Table 1. Solvent production by K. pneumoniae grown on
D-xylose (2%, w/v) under various culture conditions

Culture conditions (Headspace gas)	Optical density (540 nm)	Butanediol produced (g/L)
Aerobic	5.3	1.2
Anaerobic (N_2)	1.2	3.6
Anaerobic (N_2/CO_2, 4:1, v/v)	1.2	4.0
Anaerobic (H_2)	0.8	3.1
Anaerobic (H_2/CO_2, 4:1, v/v)	0.9	3.8
Finite air	3.7	6.2

The production of butanediol from the major sugars present in biomass hydrolyzates under finite air conditions was examined. All the sugars were utilized within two days and converted to butanediol and, to a lesser extent, ethanol (Table 2).

Table 2. Butanediol production from sugars present in biomass hydrolyzates

Substrate (20 g/L)	Optical density (540 nm)	Butanediol produced (g/L)
D-glucose	4.2	7.3
D-galactose	3.4	5.7
D-mannose	3.6	5.8
D-xylose	3.7	6.2
L-arabinose	3.6	6.0
D-cellobiose	3.7	5.9

The fermentation profiles under finite air conditions were then studied (Fig. 1). When D-glucose was used as the substrate, cell growth generally proceeded with no appreciable lag period, a pattern followed closely by sugar utilization and the decline in pH of the medium. Maximum diol and ethanol production occurred after 30 h, concurrent with the completion of growth and carbon utilization. Growth and solvent production were slower for cultures grown on D-xylose, with maximal diol production occurring after 54 h incubation.

Fig. 1. Time course of the growth and solvent production of K. pneumoniae grown under finite air conditions

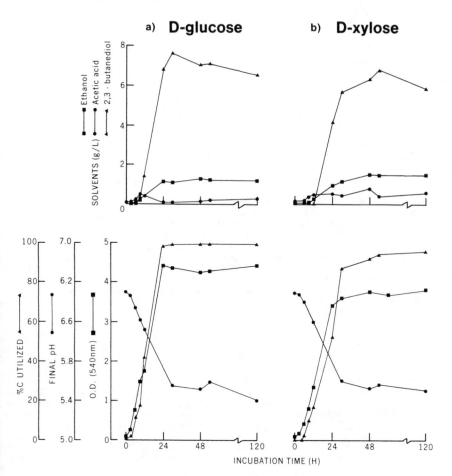

140

Subsequently, it was demonstrated that butanediol production was significantly enhanced by the presence of acetic acid in the fermentation medium (Yu and Saddler, 1982b; 1983) (Table 3).

Table 3. The effect of acetic acid addition on the
production of butanediol from D-glucose (4%, w/v)

Acetic acid added (g/L)	Sugar used (%)	Butanediol produced	
		(g/L)	(g/g)
0.0	52.0	8.0	0.38
1.0	87.0	16.4	0.47
3.0	92.0	18.4	0.50
5.0	100.0	20.0	0.50
10.0	18.0	2.5	0.14

In the presence of added acetic acid, all the major sugars known to be present in biomass hydrolyzates could be efficiently converted to butanediol at near theoretical yields (Yu et al, 1984b). The acetic acid added also resulted in the reduction of acid formation during butanediol fermentation, and eliminated the requirement for the expensive organic buffer (MES) in the medium. This resulted in a simple and inexpensive medium for large scale production of butanediol and enabled easier scale-up of the process in fermentors without the need for automatic pH control.

By adopting a combination of acclimatization and double fed-batch (daily addition of sugar and yeast extract) approaches, high butanediol levels (106 g/L and 80 g/L from D-glucose and D-xylose, respectively) could be obtained in the modified medium (Yu and Saddler, 1983). These product levels surpassed the 80-100 g/L concentrations estimated to be required for economical product recovery (Jansen, 1982). By replacing yeast extract by ammonium sulfate in the double fed-batch approach, final butanediol levels could be raised to 114.7 g/L. The process was also successfully scaled up in laboratory fermentors to give comparable product yields.

To date, these studies have demonstrated the technical feasibility of producing high levels of butanediol, at near theoretical yields from commercial sugars. The success of the process therefore depends on the availability of inexpensive substrates for the fermentation. If butanediol could be produced from lignocellulosic substrates, such as wood and agricultural residues, the commercial viability of the process would be greatly enhanced. We therefore examined the feasibility of using K. pneumoniae for the conversion of biomass to butanediol.

Since this organism could efficiently utilize all the major sugars known to be present in the hemicellulose and cellulose components of lignocellulosic substrates, we tested the possibility of establishing a single process for the direct conversion of the total carbohydrates to butanediol. Earlier we had shown that pretreatment of wood and agricultural residues by steam-explosion enhanced the enzymatic hydrolysis of the substrates (Saddler et al, 1982a; Yu et al, 1984a). Screening of the cellulolytic and xylanolytic organisms in Forintek's culture collection was then carried out (Saddler, 1982). Trichoderma harzianum was selected as the most promising source of the hydro-lytic enzymes since it produced a full spectrum of extracellular cellulase and xylanase enzymes (Saddler, 1982; Yu et al, 1984b). The enzymes present in the fungal culture filtrates could readily hydrolyze both the cellulose and hemicellulose substrates. However, the efficiency of hydrolysis declined with increasing substrate concentration (Saddler et al, 1983), presumably as a result of the accumulation of end-product inhibitors. A combined enzymatic hydrolysis and fermentation (CHF) approach was therefore studied. By carrying out the substrate hydrolysis and fermentation in a single reactor, the hydrolysis products could be continually removed by the fermentative organism and converted to fermentation products. The use of K. pneumoniae cells as the fermentative component of the CHF process has the additional advantage in that this organism could utilize all the end-products of cellulose and hemicellulose hydrolysis, viz., glucose, cellobiose, xylose, and xylobiose. Consequently the process should be more effective than comparable systems using organisms capable of converting only glucose or xylose to products.

The approach was first optimized using aspenwood xylan and solka floc as model substrates of hemicellulose and cellulose, respectively (Yu et al, 1984a; 1984b). Under the conditions established, butanediol production by the CHF approach was found to be superior to the conventional approach of sequential hydrolysis and fermentation (SHF) in both the final product levels (g/L) and the level of productivity (g/L/H). Preliminary studies using laboratory fermentors also showed that the CHF process could be easily scaled up.

The feasibility of using the CHF approach for the production of butanediol from hemicellulose and cellulose components was then investigated, with the ultimate aim of using the process for the direct conversion of combined cellulose and hemicellulose carbohydrates of biomass after a simple pretreatment such as steam-explosion (Yu et al, 1984a; 1984b). Recent studies showed that the product yields obtained from the unextracted steam-exploded substrates surpassed the combined yields of products obtained from the water-soluble hemicellulose and the water-insoluble cellulose (Yu et al, 1984b). Under the conditions used, high solvent yields (g of product per g of original untreated biomass substrate) were obtained from the steam-exploded aspenwood and agricultural residues (Table 4).

Table 4. Solvent production from steam-exploded substrates
by the CHF approach

Substrate (20 g/L)	Solvent yields (g/g of original substrate)	% Theoretical Conversion
Aspenwood (Acid-impregnated)		
220°C/20s *	14.9	40.1
220°C/40s	20.5	58.2
220°C/80s	23.0	65.4
240°C/10s	14.2	56.0
240°C/20s	10.2	34.0
Wheat Straw		
240°C/30s	19.5	52.8
240°C/60s	20.0	64.2
240°C/90s	20.6	69.4
Barley Straw		
240°C/30s	13.8	38.9
240°C/60s	13.4	43.3
240°C/90s	8.8	31.4
Corn stover		
240°C/30s	11.4	41.6
240°C/60s	9.4	39.9
240°C/90s	7.4	36.0

* Temperature and duration of steaming

The results also indicated that the optimization of butanediol production from lignocellulosic substrates required the
optimization of the fermentation process as well as the upstream
processing of the substrates, including pretreatment and
enzymatic hydrolysis. Conditions for steam-explosion would have
to be established for each type of biomass substrate in order to
maximize the carbohydrate recovery, to minimize the inhibitor
production during the steam treatment, and to maximize the enzymatic hydrolyzability (and subsequent fermentability) of the
substrates. Similarly, optimization of hydrolysis should be
investigated to establish the conditions for hydrolysis or CHF to
maximize hydrolysis efficiency and to minimize enzyme
inactivation. Continued efforts should be made to search for
enzyme preparations with more desirable characters, including a
full spectrum of activities at proper proportions for optimal
synergism, thermostability, long half-life, resistence to end-
product inhibition, etc. Although the present study demonstrated

that the fermentation conditions are well established for the production of high levels of butanediol at near theoretical yields, further improvements can be expected in the productivity (i.e., g/L/H) of the process. A thorough study should also be initiated to overcome the problem of fermentation inhibitors known to be present in steam-exploded substrates. This could be carried out by the selective removal the inhibitory materials, or by acclimatizing the organism to the inhibitors. Further work in these areas will undoubtedly improve the technical as well as economical feasibilities of converting biomass to fuels and chemicals.

ACKNOWLEDGMENT

We would like to thank H.H. Brownell, N. Levitin, K. West, and H. Normand for the preparation and analyses of lignocellulosic substrates; G. Louis-Seize for the enzyme preparations; and L. Deschatelets and J. Torrie for the fermentation and subsequent CHF studies.

REFERENCES

Dekker, R.F.H. 1983. Biotechnol. Bioeng. 25: 1127-1146.
Flickinger, M.C. 1980. Biotechnol. Bioeng. 22S: 27-48.
Jansen, N.B. 1982. Ph. D. Thesis. Purdue University, West Lafayette, Indiana.
Jansen, N.B., Flickinger, M.C., and Tsao, G.T. 1984. Biotechnol. Bioeng. 24: 362-369.
Jones, J.K.N., Purves, C.B., and Timell, T.E. 1961. Can. J. Chem. 39: 1059-1066.
Koshijima, T., Timell, T.E., and Zimbo, M. 1965. J. Polymer Sci. 11: 265-270.
Ledingham, G.A. and Neish, A.C. 1954. Ind. Ferment. 2: 27-93.
Long, S.K. and Patrick, R. 1963. Adv. Appl. Microbiol. 5: 135-155.
Lowry, O.H., Rosebrough, N.J., Farr, A.L., and Randal, R.J. 1951. J. Biol. Chem. 193: 265-275.
Mandels, M., Andreotti, R., and Roche, C. 1976. Biotechnol. Bioeng. Symp. 6: 21-34.
Miller, G.L. 1959. Anal. Chemo 31: 426-428.
Rosenberg, S.L. 1980. Enzyme Microbial Technol. 2: 185-193.
Saddler, J.N. 1982. Enzyme Microbial Technol. 4: 414-418.
Saddler, J.N., Brownell, H.H., Clermont, C.P., and Levitin, N. 1982a. Biotechnol. Bioeng. 24: 1389-1402.
Saddler, J.N., Hogan, C., Chan, M.K.-H., and Louis-Seize, G. 1982b. Can. J. Microbiol. 28: 1311-1319.
Saddler, J.N., Yu, E.K.C., Mes-Hartree, M., Levinin, N., and Brownell, H.H. 1983. Appl. Environ. Microbiol. 45: 153-160.
Shields, R. and Burnett, W. 1960. Anal. Chem. 11: 2-12.
Wentz, F.E., Marc, A.D., and Gray, M.J. 1982. J. Chromatogr. Sci. 20: 349-354.
Yu, E.K.C., Deschatelets, L., and Saddler, J.N. 1984a. Appl. Microbiol. Biotechnol. 19: 365-372.

Yu, E.K.C., Deschatelets, L., and Saddler, J.N. 1984b.
    Biotechnol. Bioeng. Symp. 14 (in press).
Yu, E.K.C. and Saddler, J.N. 1982a. Biotechnol. Lett. 4: 121-126.
Yu, E.K.C. and Saddler, J.N. 1982b. Appl. Environ. Microbiol.
    44: 777-784.
Yu, E.K.C. and Saddler, J.N. 1983. Appl. Environ. Microbiol.
    46: 630-635.

OPTIMIZATION OF 2,3 BUTANEDIOL PRODUCTION FROM GLUCOSE IN BATCH AND FED-BATCH CULTURES OF BACILLUS POLYMYXA

V.M. Laube, D. Groleau, S.M. Martin and I.J. McDonald

Division of Biological Sciences
National Research Council
Ottawa, Canada

Abstract

The fermentation of glucose (5%) to 2,3-butanediol by Bacillus polymyxa was optimal at 1.5% yeast extract. At this concentration of yeast extract, the diol production was increased when the glucose concentration was raised to 7%. At 8% glucose and higher, decreased diol yields occurred. In fed-batch cultures up to 10% glucose was readily utilized without inhibition of diol production although yields were not higher than in 7% glucose batch cultures.

Introduction

Yeast extract is known to stimulate 2,3-butanediol production from glucose by Bacillus polymyxa (Laube et al., 1984b). To assess the effect of substrate concentration in the presence of optimum amounts of yeast extract, 2,3-diol production was determined in batch cultures with increased amounts of glucose. In addition, the effects of increased

substrate on solvent production in fed-batch cultures, to which
substrate was added at intervals during the fermentation, were
examined.

Materials and Methods

Bacillus polymyxa (NRCC 9035), the organism used in
these studies, was maintained as previously described (Laube et
al., 1984a). Inocula and all experimental cultures were grown
in 125 mL Erlenmeyer flasks using 50 mL of the basal medium of
Laube et al. (1984a) with specified concentrations of yeast
extract. The substrate, glucose, prepared as a 50% (w/v)
solution (pH 7.2), was added to the sterile basal medium, which
was prepared concentrated, to achieve a final 5-10% (w/v)
substrate level. For fed-batch cultures, additional glucose
(2%, w/v), with and without 1.5% yeast extract, was added at
designated time intervals. Batch cultures were analyzed at zero
time and at designated intervals; fed-batch cultures were
analyzed at zero time and before and after glucose additions.

Aliquots (1 ml) were removed for analysis and, after
pH values were recorded, they were centrifuged for 4 min. at
13-15,000 × g. The supernatant fluids were analyzed for
2,3-diol, acetoin and ethanol using gas chromatography (Ackman,
1972) with 1,2-butanediol as internal standard and for glucose
by the method of Miller (1959) with corrections made for acetoin
(Murphy et al., 1951). All cultures were incubated at 30°C with
shaking at 125 rpm.

Results and Discussion

Yeast extract has been shown to have a marked effect

on diol production and glucose utilization with B. polymyxa
(Laube et al., 1984b). Increasing the concentration of yeast
extract in medium containing 5% glucose resulted in increased
diol yields; an optimal value of 18 g/L was achieved with 1.5%
yeast extract (Figure 1). Rates of glucose utilization also
increased with an increase in yeast extract concentration with
the best rates being achieved at 1.5 and 2%.

The pH values of all cultures initially fell from pH
6.5-6.7 to pH 5.0-5.1. This drop was followed by a rise in pH
in those cultures containing 1.0 and 1.5% yeast extract and
coincided with the depletion of substrate and with a decrease in
diol levels. This decrease in diol was accompanied by an
increase in the acetoin levels with the highest level of 12.1
g/L being reached with 1.5% yeast extract at day 4. Ethanol

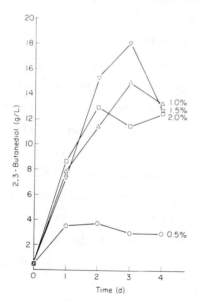

Figure 1: The effect of yeast extract on 2,3-butanediol
production in cultures containing 5% glucose.

values, although low in all experiments, did increase progressively from 0.6 g/L to 3.8 g/L with an increase in yeast extract from 0.5 to 2%.

Although the best diol yields in cultures containing 5% glucose were achieved with 1.5% yeast extract, diol recovery under these conditions was only 42% of theoretical. An attempt was thus made to improve yields and recovery by increasing the glucose concentration.

Cultures containing 5 and 6% glucose showed highest initial rates of diol production (Figure 2A). However, highest diol yields (28 g/L) and recovery rates (80% of theoretical) were achieved in cultures containing 7% glucose. Rates of diol production with 8, 9 and 10% glucose leveled off after 2 days at 13 g/L with 8 and 9% and 8 g/L with 10% substrate. All cultures, with the exception of the one containing the highest substrate, showed similar rates of glucose utilization (Figure 2B). By day 4, all the glucose was used in cultures containing 5, 6 and 7% substrate; least glucose was used with 10% sugar. Acetoin levels, initially low, increased once all of the glucose was metabolized and when diol levels began to fall. This increase in acetoin was accompanied by a rise in pH from 5, the value to which the pH initially fell from pH 6.7 in all cultures at the commencement of the fermentation. Since diol levels had not decreased by day 4 even though all of the substrate was used, the pH did not increase in cultures containing 7% glucose. Ethanol values ranged from 2 to 3 g/L with no obvious trend evident.

149

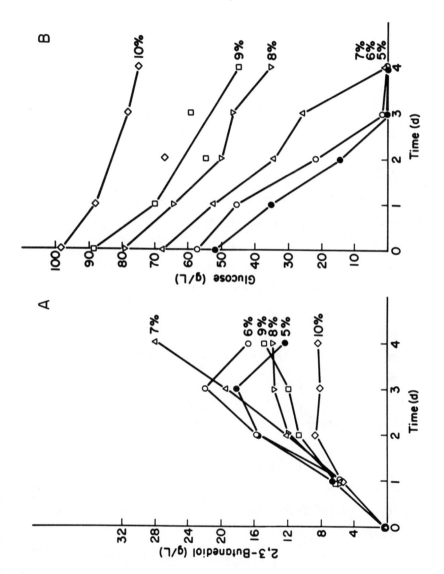

Figure 2.   Profiles of (A) 2,3-butanediol production and (B)
glucose utilization in cultures containing 1.5% yeast
extract and various glucose concentrations.

The above results suggest that initially high concentrations of substrate in batch cultures of B. polymyxa may be inhibitory to diol production. However, Yu and Saddler (1981) have shown that fed-batch cultures of Klebsiella pneumoniae use significantly more substrate and produce greater amounts of solvent (diol and acetoin) than batch cultures. Therefore, the effects of the addition of substrate in fed-batch cultures of B. polymyxa were studied.

Glucose was added (20 g/L) every 2 days to cultures initially containing 50 g/L (Figure 3). The rates of substrate utilization up to day 6 were similar after which time a decreased rate was observed. Diol yields and rates were comparable to a batch-culture containing 7% glucose. However, considering the amount of substrate used, yields were considerably lower. At day 6, yields were 60% whereas by day 8 they were only 50% of theoretical. Although it has been reported that B. polymyxa is not affected by such levels of diol (Seri report, 1981) and can tolerate up to 10% (w/v), it is possible that this strain of the bacterium is more sensitive to diol.

Even though substrate was not limiting and no overt decreases in diol levels were observed, a substantial increase in acetoin levels occurred by day 6 and by day 8, 14.6 g/L were present. In addition, the pH of these cultures fell during the initial stages of the fermentation and then rose to pH 5.5 by day 4 and then leveled-off at 5.8 by day 6. Ethanol levels were between 2 and 3 g/L and no trend was observed.

Similar results were obtained in fed-batch experiments when a mixture of glucose (20 g/L) and yeast extract (15 g/L)

was added thus indicating that yeast extract was not the
limiting factor in diol production.

Since glucose levels tended to drop to 10 g/L when
cultures were fed every 2 days, cultures were established with
daily feedings starting after the initial 2 day fermentation.
In these experiments, glucose levels rarely fell below 20 g/L.
The trends of all the parameters observed were the same as in
the previous experiments. The same rate of diol production was
observed and no more diol was produced. However, since less
glucose was used during the fermentation, a 70% yield (of
theoretical) was obtained by day 6.

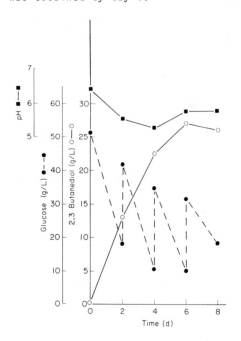

Figure 3:   Profiles of 2,3-butanediol levels, glucose levels and
            pH in fed-batch cultures containing 1.5% yeast
            extract.

References

Ackman, R.G. (1972) J. Chromatogr. Sci. 10: 560-565.

Laube, V.M., D. Groleau and S.M. Martin (1984a) Biotechnol. Letters 6: 257-262.

Laube, V.M., D. Groleau and S.M. Martin (1984b) Biotechnol. Letters 8: 535-540.

Miller, G.L. (1959) Anal. Chem. 31: 426-428.

Murphy, D., D.W. Stranks and G.W. Harmson (1951) Can. J. Technol. 29: 131-143.

Seri (Solar Energy Research Institute) Report (1981). Alcohol Fuels Program Technical Review, Winter, 1981.

Yu, E.K.C. and J.N. Saddler (1983) Appl. Environ. Microbiol. 46: 630-635.

Biological Production of Economically-Recoverable
Products from Dilute Ethanol Streams

David W. Armstrong,[1,2] Stanley M. Martin[1] and Hiroshi Yamazaki.[2]

Division of Biological Sciences, National Research Council of Canada, Ottawa, K1A OR6,[1] Department of Biology, Carleton University, Ottawa, K1S 5B6,[2] Canada.

One area of growing interest related to biotechnological processes is the recovery of low concentration ethanol for which conventional distillation processes are not economical or feasible. Several existing fermentations or many under consideration (e.g. lignocellulosic conversion) produce dilute ethanol streams which are not amenable to conventional recovery techniques. Various physical/chemical methods are under consideration however technical problems currently limit their application. A novel approach to separate ethanol from dilute alcohol streams would be to biologically convert the ethanol into more readily recovered products such as esters or aldehydes useful as fuels/chemical feedstocks. At the same time an upgrading in unit value of the alcohol would result. We have found that the yeast Candida utilis can convert ethanol to either ethyl acetate or acetaldehyde (both useful as fuels/chemical feedstocks). The concept of a biological-based means to recover low concentration ethanol and a model system using C. utilis are discussed.

Recovery of Dilute Ethanol

One area of growing interest related to biotechnological processes is the recovery of ethanol from dilute solutions for which conventional distillation processes are not economical or feasible. Conventional distillation recovery processes are economically suited for higher alcohol concentrations (typically 6% or greater) (Fig. 1). However several fermentations produce dilute ethanol streams which are not amenable to conventional recovery methods. For example, the brewing and distilling industries generate large volumes of process wastes containing dilute ethanol. Fermentation of cellulose and hemicellulose, currently under consideration, generally also produce low concentration ethanol streams (Table 1). Therefore, processes which are less energy intensive and less costly than conventional distillation are being sought (Table 2). However technical problems currently limit their application.

The conversion of ethanol to products (e.g. esters or aldehydes) which are suitable for use as fuels and/or chemical feedstocks and which are recoverable by low energy separation techniques (e.g. solvent extraction, low temperature evaporation) would be of great benefit. As well a concurrent upgrading in unit value of the ethanol would result.

The distribution coefficients of many esters, ethyl acetate for example, are orders of magnitude greater than ethanol in numerous solvents including unleaded gasoline (Table 3). Thus simple solvent extraction may be used. It is possible that a 0.5% aqueous solution of these esters could be concentrated 5-10 fold into gasoline layer by simple solvent extraction (Datta, 1981). Many of these esters, apart from current use as solvents have potential for use as octane-enhancing gasoline additives (Table 4) and in production of ketones, ethers, etc.

154

Table 1.   Microbial Production of Ethanol

Microorganism (type)	Process Details	Substrate (conc. %, w/v)	Ethanol Produced	Reference
Pachysolen tannophilus (yeast)	Batch/aerobic	xylose[a] 7.0 4.3	0.7 2.0	Detroy et al., 1982
Acetivibrio cellulolyticus (bacterium)	Batch/ anaerobic	cellulose 10.0	0.1	Armstrong et al., 1983
Zymomonas mobilis (bacterium)	Continuous/ Immobilized/ anaerobic	glucose 10	4.5	McGhee et al., 1982
Saccharomyces cerevisiae (yeast)	Continuous/ Immobilized/ anaerobic	glucose 10	5.0	McGhee et al., 1982

[a]Predominant (C_5) monomer from hemicellulose hydrolysis.

Table 2.   Alternative Ethanol Separation Processes[a]

Process	Ethanol Productivity	Comments
Direct extractive fermentation	Potentially very high	No suitable extractant.
Selective membrane fermentation	Potentially very high	Membranes with different throughput rates have not yet been developed.
Vacuum fermentation	Very high	Mechanically complicated equipment. Contamination problems. Pure oxygen must be sparged.
Adsorption/ desorption	Potentially high	Energy requirements unknown. Need ethanol conc. 10% or greater
Electrochemical oxidation	Very high	Catalyst poisoning by many fermentation by-products. Energy input significant.

[a]Table adapted from Maiorella et al. (1981).

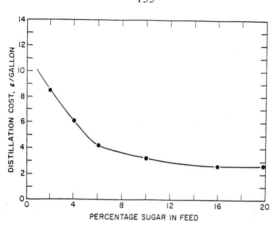

Figure 1. Economics of Ethanol Distillation.

    Conversion of ethanol to aldehydes such as acetaldehyde may also be of economic advantage. Several aldehydes have very low boiling points (e.g. acetaldehyde, $21^{o}C$). Acetaldehyde may be stored, once evaporated from the fermentation broth, in a polymerized form such as paraldehyde (b.p $124^{o}C$). Both acetaldehyde and paraldehyde have been suggested as diesel fuel substitutes (increasing the cetane number). Acetaldehyde can be used as an on-site gas turbine fuel substitute

Table 3.

Solvent Extraction of Organic Acid Esters

	Extraction coefficient (K_D) of:	
Solvent	Ethyl Acetate	Ethyl Butyrate
Benzaldehyde	12.4	160
1-octanol	4.9	52.5
unleaded gasoline	4.2	68.5

From Datta (1981)

(Meshbesher, 1982) or a space heating fuel. Acetaldehyde is miscible with standard fuels such as fuel oil and reportedly promotes combustion. Many useful chemicals such as butanol, butadiene can be produced from acetaldehyde via simple chemical processing techniques.

Table 4.
Octane Indices of Various Esters

	BPT (°C)	Octane Index (Research)
Ethyl acetate	77.2	∿ 117
Ethyl propionate	99.1	∿ 118
Ethanol	78.3	110-125

Classical Production of Esters/Aldehydes from Ethanol

The production of esters/aldehydes could be of economic importance to recover low concentration ethanol. Unfortunately conventional synthetic routes to produce either esters or aldehydes from dilute alcohol streams would not be economic or feasible in many cases.

The reaction of carboxylic acids with alcohol to produce esters is a well known process. However highly concentrated alcohol and organic acid solutions would be required.

Numerous classical methods for production of acetaldehyde from ethanol involving elevated temperatures and certain catalysts have been used. Recently interest has been developing in the conversion of fermentation ethanol to acetaldehyde by electrochemical oxidation. In this conversion many of the by-product of a typical ethanolic fermentation such as acetate, proteinaceous substances and even the intended product - acetaldehyde could act as catalyst poisons and would quickly curtail the operation of the fuel cell. Considerable energy input would also be required.

Biological Production of Esters/Aldehydes from Fermentation Ethanol

A novel approach to deal with dilute ethanol streams would be to biologically convert the alcohol to either an ester or an aldehyde. A biological means to separate these dilute alcohol streams into more economically-recoverable products would be compatible with fermentations since they would operate under similar conditions (temperature, pressure, etc.).

Various yeast of the genera Saccharomyces and Hansenula produce ethyl acetate from glucose and/or ethanol. Saccharomyces sp. produce very low concentrations of this ester (i.e. levels found in alcoholic beverages) whereas Hansenula anomala produces significant amounts of the ester from glucose or ethanol but only after extensive fermentation (several weeks).

We have found that iron-limited <u>Candida</u> <u>utilis</u> grown on a glucose-containing
dium produced significant levels of ethyl acetate (Armstrong et al., 1984a)
ig. 2). A low level of aeration was chosen to allow delineation of separate phases.
hyl acetate accumulated in cultures without iron supplementation but not in
on-supplemented cultures. In both cases ethanol and the cell mass density
creased until glucose was depleted (2 d) and the yield of ethanol was similar
bout 90% of theoretical yield). In the culture without iron supplementation, the
lls began to utilize ethanol after a 2 d lag period and accumulated acetic acid
d cell mass. Following this, ethyl acetate accumulation began and continued while
etic acid declined. Clearly both acetic acid and ethyl acetate resulted from
hanol utilization.

The effect of ethanol concentration on the efficiency of conversion to ethyl
:etate was studied (Fig. 3). Ethanol-adapted cells were resuspended in nitrogen-
·ee medium containing various concentrations of ethanol. The rate of conversion of
:hanol to ester was proportional to the level of the alcohol up to about 10 g/L
·ring the first 5 h of incubation. Ethanol concentrations greater than about 10 g/L,
·ring this same time period, showed some inhibitory effects which became more
·ident with increases in the level of ethanol. However, by 24 h, the yield
·fficiencies (dashed line) were similar to those found at the 10 g/L level. Thus
·e most rapid conversion of ethanol to ethyl acetate occurs at about 10 g/L. Ethanol
·ncentrations greater than 35 g/L were strongly inhibitory to ester accumulation.
: was found that at ethanol levels greater than about 35 g/L that there was
·gnificant accumulation of another volatile product - subsequently identified by
:/MS as acetaldehyde (Armstrong et al., 1984b).

The product acetaldehyde has a relatively low boiling point of 21°C and
herefore would allow for simple evaporative techniques for its recovery from
·rmentation broths. Figure 4 shows the change in product distribution from ethyl
·cetate to acetaldehyde depending upon the concentration of ethanol. Although at all
·vels of ethanol the rate of its utilization is similar the products accumulating
·re quite different. Ethanol concentrations of about 35 g/L or less favour ethyl
·cetate (Fig. 4a). At an intermediate level of ethanol (Fig. 4b) acetaldehyde was
·he predominant product accumulating up until the level of ethanol declined below
·a. 35 g/L. At higher levels of ethanol (Fig. 4 c), acetaldehyde continued to
·ccumulate. Acetic acid was found to accumulate from the further oxidation of
·cetaldehyde under the conditions used ('closed system').

The effect of ethanol levels on acetaldehyde yields was studied. Acetaldehyde
·ields were proportional to ethanol levels from about 35 to 60 g/L. A decline in
·ccumulation and yield was seen progressing to higher alcohol levels (Fig. 5).
It has been speculated that in C. utilis, ethyl acetate is formed from the react-
·on of acetyl-CoA with ethanol (Thomas and Dawson, 1978). Acetaldehyde is known to
·nhibit acetyl-CoA synthetase which catalyses the formation of acetyl-CoA from
·cetic acid. The results suggest that higher levels of ethanol cause accumulation
·f higher intracellular levels of acetaldehyde which in turn inhibits acetyl-CoA
·ormation thereby reducing ethyl acetate accumulation.

The present results were obtained with cells suspended in medium in sealed
·ntainers ('closed systems'). It is noted that acetaldehyde can be further oxidized
· acetic acid. Preliminary experiments in an open system have indicated that
·ntinuous removal of acetaldehyde ('open system'), minimizes accumulation of acetic
·id and allows for production of significant levels of acetaldehyde.

Collectively the results have demonstrated the feasibility for the conversion of
·lute ethanol streams to more economically-recoverable products. Other
·croorganisms may have potential to be used directly or through physiological/
·netic manipulation for conversion of ethanol to various esters and aldehydes
·itable for use as fuels/chemical feedstocks.

158

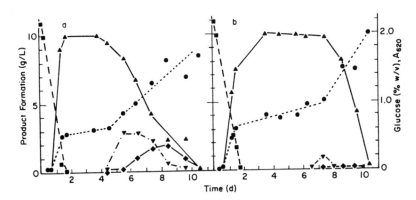

Figure 2. Accumulation of ethanol, acetic acid, ethyl acetate, and cell mass in Candida utilis. Candida utilis was grown in a minimal salts medium containing 20 g/L glucose (a) without or (b) with FeCl (100μ M). The medium (100 mL) was inoculated with 1 mL of an overnight culture and shaken at 150 rpm in a 250-mL Wheaton serum bottle. Initial pH 5.8 dropped to 2.3 after 24 h. The symbols refer to (▲) ethanol, (▼) acetic acid (shown 10 X actual concentration), (X) ethyl acetate, (■) glucose, and (●) A_{620}.

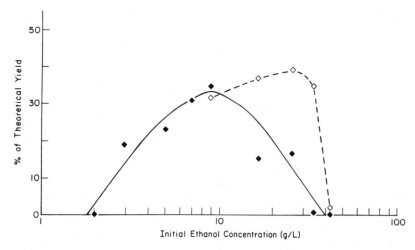

Figure 3. Effect of ethanol concentration on ester yield. Ethanol-adapted cultures were isolated by centrifugation and resuspended (A_{620} = 0.55) in 10 mL nitrogen-free medium (in 160-mL Wheaton vials). Ethanol added at ca. 2, 4, 6, 8, 10, 20, 30, 40, and 50 g/L. Percent of theoretical yield of ethyl acetate from ethanol is shown at (◆) 5 h or (◇) 24 h.

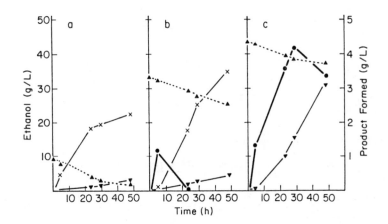

Figure 4. Accumulation of products at different levels of ethanol. Ethanol-adapted C. utilis cells were washed and resuspended to A620 = 2.0 in the medium (pH 6.0) containing ethanol at about 10 g/L (part a) 35 g/L (part b) and 45 g/L (part c). (▲) ethanol; (X) ethyl acetate; (▼) acetic acid; (●) acetaldehyde.

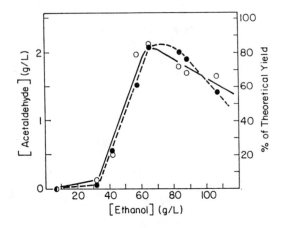

Figure 5. Effect of ethanol concentration on acetaldehyde accumulation of yield. Ethanol-adapted C. utilis cells were resuspended to A620 = 2.0 in the nitrogen-free medium (pH 7.0) containing different levels of ethanol and incubated for 5 h. Theoretical yields of acetaldehyde were based on the conversion of 1 mole of ethanol to 1 mole acetaldehyde (● ⸺ ●) acetaldehyde accumulation; (o — o) yield.

160

References

Armstrong, D.W., S.M. Martin, H. Yamazaki. (1983). Bacterial Fermentation of Cellulose: Effect of Physical and Chemical Parameters. Biotechnol. Bioeng. 25, 2567-2575.

Armstrong, D.W., S.M. Martin, H. Yamazaki. (1984a). Production of Ethyl Acetate from Dilute Ethanol Solutions by Candida utilis. Biotechnol. Bioeng. 26, 1038-1041.

Armstrong, D.W., S.M. Martin, H. Yamazaki. (1984b). Production of Acetaldehyde from Ethanol by Candida utilis. Biotechnol. Lett. 6, 183-188.

Cysewski, G.R., C.R. Wilke. (1978). Process Design and Economic Studies of Alternative Fermentation Methods for the Production of Ethanol. Biotechnol. Bioeng. 1421-1445.

Datta, R. (1981). Production of Organic Acid Esters from Biomass - Novel Processes and Concepts. Biotechnol. Bioeng. Symp. No. 11, 521-532.

Detroy, R.W., R.L. Cunningham, R.J. Bothast, M.O. Bagby, A. Herman. (1982). Bioconversion of Wheat Straw Cellulose/Hemicellulose to Ethanol by Saccharomyces uvarum and Pachysolen tannophilus. Biotechnol. Bioeng. 24, 1105-1113.

Maiorella, B., Ch.R. Wilke, H.W. Blanch. (1981). Alcohol Production and Recovery. Adv. Biochem. Eng. 20, 43-92.

McGhee, J.E., G. St. Julian, R.W. Detroy, R.J. Bothast. (1982). Ethanol Production by Immobilized Saccharomyces cerevisiae, Saccharomyces uvarum, and Zymomonas mobilis. Biotechnol. Bioeng. 24, 1155-1164.

Meshbesher, T.M. (1982). Method for Making Acetaldehyde from Ethanol. U.S. Patent No. 4,347,109.

Thomas, K.C., P.S.S. Dawson. (1978). Relationship Between Iron-limited Growth and Energy Limitation During Phased Cultivation of Candida utilis. Can. J. Microbiol. 24, 440-447.

RECENT PROGRESS IN OBTAINING ETHANOL FROM XYLOSE

H. Schneider, A.P. James, J. Labelle, H. Lee,
G. Mahmourides, R. Maleszka
National Research Council

and

G. Calleja and S. Levy-Rick
Ethanol from Cellulose Program

ABSTRACT

Considerable progress has been made recently in using yeasts to obtain ethanol from xylose and from a lignocellulosic hydrolyzate. The developments have occurred at both the basic and applied level.

At the basic level, insight has been obtained into physiological aspects of yeast cultures on xylose under aerobic conditions that are relevant to their control so that optimum ethanol is obtained. Aerobic batch cultures of Pachysolen tannophilus and of Candida tropicalis were found to undergo a transition from an oxidative to a fermentative state. Ethanol accumulated only during the fermentative state. While in this state, ethanol accumulated when the concentration of dissolved oxygen was either low or high, suggesting that the major effect of aeration rate on ethanol yield was the result of events that took place during the preceeding oxidative state. The factors responsible for initiating the transition to the fermentative state may not be identical to those which trigger an analogous transition in brewing yeasts on glucose. An additional development at the basic level has been the isolation of mutants of Pa. tannophilus that produce more ethanol from xylose than the wild-type.

At the applied level, work carried out in the course of the Ethanol from Cellulose Program has resulted in the ability to obtain 5% w/v ethanol from 10% w/v xylose within 24 hours. In addition, substantial yields of ethanol could be obtained with a hydrolyzate of poplar prepared using hydrofluoric acid. A concentration of 5% w/v was obtained when the sugar concentration was 12% w/v.

INTRODUCTION

A number of yeast species have been identified that convert xylose to ethanol (e.g. 1-4). However, methodology suitable for their use with industrial substrates has still to be developed. Two general reasons are responsible for this situation. One is the result of our poor understanding of the physiology underlying the growth of yeasts on xylose and of its fermentation. For example, ethanol yields and the production of undesirable by-products vary widely with aeration rate (e.g. 5,6) but there is no rational basis on which to choose or to vary aeration conditions to produce the optimum yield of ethanol. The other reason is the compounds present in hydrolyzates that inhibit the fermentation of xylose. Significant yields of ethanol can be obtained only after extensive pretreatment for their removal (e.g. ion-exchange (2)). Inexpensive procedures to obviate the effects of inhibitors that are sufficiently inexpensive for use at the industrial level do not seem to be available.

The present report provides an overview of progress made recently in using yeasts to obtain ethanol with high efficiency from pure xylose or from a lignocellulosic hydrolyzate.

RESULTS AND DISCUSSION

Physiological Aspects of Yeast Cultures on Xylose

A finding that is relevant for the development of efficient aerobic fermentations concerns a change in the ability to use oxygen that can occur in cultures of Pachysolen tannophilus and of Candida tropicalis (7). As cultures of these organisms age, they pass from an oxidative to a fermentative phase. The significance of this transition with respect to ethanol accumulation is that it occurs only during the fermentative phase. An additional significant factor is that ethanol can be accumulated during the fermentative phase when the concentration of dissolved oxygen is either low or high, suggesting that aeration rate effects on ethanol yield may be largely the result of events that occur during the oxidative phase.

The transition was found by following the time course of the concentration of dissolved oxygen under appropriate conditions of aeration (Figure 1). With cultures inoculated initially at a low cell density, the concentration of dissolved oxygen decreased initially as the result of the proliferation of cells that used oxygen rapidly. However, the capacity to use oxygen eventually decreased, as shown by the increase in the concentration of dissolved oxygen. The changes in capacity to use oxygen indicated by the time course for dissolved oxygen was confirmed by measurements of the specific rate of oxygen uptake of cells isolated from the culture at intervals before and after the transition (Figure 2). Confirmation was obtained as well in a bioreactor where the specific and volumetric rate of oxygen use was measured (Table 1).

Because ethanol was accumulated largely during the period when the capacity of the culture to use oxygen was low, that is during the fermentative phase, maintaining cells continuously in this phase would be expected to enhance yield. However, it is not yet known how to lock cells into the

163

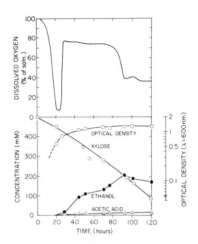

FIGURE 1.   Variation in Dissolved Oxygen Concentration and Other Variables
Under Constant Aeration After Inoculation of <u>Pa. tannophilus</u> NRRL
Y2460 on 500 mM D-xylose in 0.67% Yeast Nitrogen Base.   The
bioreactor consisted of a loosely-capped 500 ml erlenmeyer flask.
It was filled with 100 mls of medium and shaken on a gyratory shaker
at 200 rpm at 30°C.

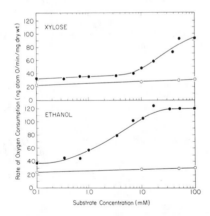

FIGURE 2.   Stimulation of the Specific Rate of Consumption of Oxygen by
D-xylose and by Ethanol in Cells Isolated from a Culture Before and
After the Transition had Occurred.   The culture from which the cells
were isolated was a replicate of that of Figure 1 except that 100 mM
xylose was used.   Cells isolated 24 hours after inoculation (filled
circles) and 48 hours after (open circles) represent those before
and after occurrence of the transition, respectively.   The rate with
0.1 mM substrate with cells isolated after the transition was eqaul
to the endogenous value.

TABLE 1

Rate of Oxygen Consumption and Other Variables in a Culture of
Pa. tannophilus Before and After the Transition in a Bioreactor
Run at a Constant Rate of Aeration*

Variable	Before Transition	After Transition
Time after inoculation (hours)	19	30
Cell dry weight (mg/ml)	2.6	5.5
Oxygen in medium (% of saturation)	6.25	78.0
Oxygen in head space (% v/v)	4	16.25
Specific rate of O_2 consumption (mmol/min/mg dry wt cells $\times 10^5$)	7.15	0.95
Volumetric rate of O_2 consumption (mmol/min/liter $\times 10^4$)	1.86	0.52
Ethanol (% w/v)	0.05	0.39

*The bioreactor was a Model KLF 2000 Fermenter (Bioengineering AG, Wald
Switzerland), which had a capacity of 2.4 liters. The volume of medium used
was 1.3 liters and consisted of 266 mM xylose in 0.67% yeast nitrogen base.
The rate of aeration was 0.026 VVM (34 mls/min) and the stirring speed was
2000 rpm. The optical density on inoculation was 0.01.

fermentative phase, since the factors responsible for its induction have not
been identified. They do not seem to be identical to those that cause the
Crabtree effect, a phenomenon whereby oxidative processes in S. cerevisiae are
inhibited by the presence of D-glucose. The Crabtree effect is induced
rapidly, within 1 hour. In addition, it is induced by the presence of
relatively low concentrations of sugar, 0.5 mM (8). In contrast, induction of
the fermentative state by xylose did not occur until about 24 hours after
exposure to an appreciably higher concentration of sugar, ~100 to 500 mM.
Possibly, the transition was the result of events that occurred during the
initial stages of the culture.

Mutants that Accumulate more Ethanol

One of the factors that may contribute to the low yield of ethanol
obtained under some aerobic conditions is the ability of some yeasts to oxidize
ethanol even in the presence of considerable amounts of D-xylose (9). In
addition it is possible for ethanol to be produced and to be used concurrently
when the carbon source is D-xylose (9). This latter phenomenon might
contribute to the lack of accumulation of ethanol at some stages of the
culture. In an approach to obtain strains that allow the accumulation of
enhanced amounts of ethanol, mutants were sought that could not use ethanol for

growth. The approach was successful. Twelve loci were found which influenced growth on ethanol, and mutations in three resulted in strains that produced more ethanol than the wild-type (Table 2) (10). The increases can exceed 50% of the value found with the wild-type in aerobic batch cultures. The development of a genetic mating system for Pa. tannophilus (11) and the enhanced ethanol yields produced by polyploids (12) suggests that the ethanol-negative mutants could be used to construct even more efficient fermenters.

Ethanol from Cellulose Program

An objective of the Ethanol from Cellulose Program is to evaluate the use of hydrofluoric acid in hydrolyzing lignocellulosics. Some of the work carried out in connection with this program was concerned with obtaining ethanol from the sugars present in such hydrolyzates.

In the course of the program, it was found to be possible to obtain 5% w/v ethanol from 10% w/v xylose in about 24 hours. This yield, 0.5 g ethanol/g xylose, is close to a theoretical maximum, 0.51 g/g.

Using a hydrolyzate of poplar that was prepared with hydrofluoric acid, concentrations of 5% w/v ethanol have been obtained when the sugar concentration was 12% w/v. This yield corresponds to 83% of theoretical. The pretreatment used was chosen with a view to cost-effectiveness.

TABLE 2

Maximum Concentration of Ethanol Accumulated on 4% Xylose

Strain	Ethanol % w/v
Y2460 (wild-type)	0.65
P119-6D	0.77
P139-16D	1.0
P163-16C	0.99

REFERENCES

1. Maleszka, R., and Schneider, H. (1982). Can. J. Microbiol. 28, 360-363.

2. Gong, C.-S., and Tsao, G.T. (1983). Royal Society of Canada International Symposium on Ethanol from Biomass, pp. 525-556, H.E. Duckworth and E.A. Thompson, editors.

3. Toivola, A., Yarrow, D., van den Bosch, E., van Dijken, J., and Scheffers, W.A. (1984). Appl. Environ. Microbiol. 47, 1221-1223.

4. Margaritis, A., and Bajpai, P. (1982). Appl. and Environ. Microbiol. <u>44</u>, 1039-1041.

5. Schneider, H., Wang, P.Y., and Chan, Y.K. (1981). Biotechnology Letters, <u>3</u>, 89-92.

6. du Preez, J.C., and van der Walt, J.P. (1983). Biotechnology Letters, <u>5</u>, 357-362.

7. Mahmourides, G., Lee, H., Maki, N., and Schneider, H. (1984). Biotechnology (in press).

8. Woehrer, W., and Roehr, M. (1981). Biotechnol. Bioeng. <u>23</u>, 567-581.

9. Maleszka, R. and Schneider, H. (1982). Appl. Environ. Microbiol. <u>44</u>, 909-912.

10. Lee, H., James, A.P., Zahab, D.M., Mahmourides, G., Maleszka, R., and Schneider, H. Sixth Symposium on Biotechnology for Fuels and Chemicals, Gatlinburg, Tenn., May 15-18, 1984.

11. James, A.P. and Zahab, D. (1982). J. Gen. Microbiol. <u>128</u>, 2297-2301.

12. Maleszka, R., James, A.P., and Schneider, H. (1982). J. Gen. Microbiol. <u>128</u>, 2495-2500.

Nutritional requirements of *Clostridium acetobutylicum* ATCC#824

F.W. Welsh, I.A. Veliky

National Research Council, Division of Biological Sciences
Ottawa, Canada, K1A ØR6

ABSTRACT

Physiological characteristics and biosynthetic pathways of microbial cells can be altered and optimized by modification of the micro- and macro- environments of the cells. Such manipulation can be achieved either by changing some of the physical parameters of the fermentation (e.g. temperature, pH, agitation rate, gasification, etc.) or by modifying the chemical composition of the nutrient medium. The latter approach was used as the initial step to determine the role of nitrogen, phosphate and calcium and the effect of their interactions on solvent production by *Clostridium acetobutylicum* ATCC#824. These experiments were performed in a defined synthetic medium.

Several organic and inorganic nitrogen sources were tested as alternatives to natural protein, non-defined amino acid and peptide mixtures or to ammonium acetate. Total solvent production and reaction rates were increased by using a combination of glutamic acid and ammonium chloride as the nitrogen source. Optimum growth rate and solvent production were obtained with 20 to 30 mM glutamic acid and 2.5 to 5.0 mM ammonium chloride in the nutrient medium. These combinations resulted in the production of 1.0 to 1.2% (w/v) butanol with 1.6 to 1.9% (w/v) total solvent while leaving 0.1 to 0.2% (w/v) residual butyric acid in the medium.

The interaction of calcium and phosphate was examined and interpreted by changes in solvent production when calcium and phosphate concentrations were changed. Phosphate at concentrations of 10 mM to 30 mM, without calcium addition, resulted in optimum solvent production (1.2% w/v total solvents). The addition of 2.5 mM to 5.0 mM calcium with 30 mM to 70 mM phosphate increased solvent production while inhibitory effects on solvent production were detected when 20 mM calcium concentrations were used.

INTRODUCTION

The medium composition necessary to maintain the metabolic functions of <u>Clostridium</u> at their potential maximum while reducing fermentation time has been of concern throughout the history of the acetone–butanol fermentation. Attempts to optimize the fermentation have traditionally been performed in complex media with undefined components. Recently, some of the major nutrient components have been defined and optimized in a synthetic medium using batch and continuous fermentations (1, 2). The present paper presents some results of studies on the metabolic activities and nutritional requirements of <u>Clostridium acetobutylicum</u> ATCC #824 . Specifically, it deals with the effect of nitrogen source and its concentration on cell growth and solvent production . Preliminary studies have been performed on calcium and phosphate requirements and their interaction . The results demonstrate the effect of calcium and phosphate interaction, based on solvent production, and indicate their optimum concentration range in a defined, synthetic medium in batch culture.

MATERIALS AND METHODS

INOCULUM: *Clostridium acetobutylicum* ATCC #824 was maintained on RCM at 37°C under anaerobic conditions. This culture (7 days old) was then used as inoculum (10% v/v) into a medium consisting of (per 1 Liter): KH_2PO_4, 1.50 g; K_2HPO_4, 3.00 g; $CaCl_2.2H_2O$, 1.50g; NaCl, 1.00 g; $FeSO_4.7H_2O$, 1.25 mg ; $(NH_4)_2SO_4$, 1.00 g; $MgSO_4.7H_2O$, 0.20 g; $MnSO_4.H_2O$, 2.80 mg ; yeast extract, 5.00 g and glucose, 50.00 g. After seven days of cultivation in anaerobic stationary conditions at 37°C this culture was transferred into a fresh medium of the same composition to prepare a subculture. A 2% (v/v) aliquot of a 72 hour old subculture was used as an inoculum for the small scale nitrogen experiments and the calcium and phosphate studies. An aliquot (3.6%) of a 24 hour old subculture was used as inoculum for batch nitrogen fermentation studies.

SMALL SCALE NITROGEN STUDIES: Ammonium chloride, ammonium sulfate, ammonium acetate, glutamic acid, aspartic acid, alanine and lysine were tested as nitrogen sources in a defined base medium (Table 1). pH was adjusted to 6.0 before autoclaving. Vitamins were filter sterilized and added after autoclaving. In the initial studies all nitrogen sources were added to give a final concentration of 30 mM amino or ammonium nitrogen. Total experimental volume of medium was 25 mL in 50 mL vials. The vials were incubated at 37°C under anaerobic conditions. Samples were taken after 72 hours and 144 hours, and analysed by the subsequently described methods.

NITROGEN FERMENTATION STUDIES : These studies were carried out in V–fermentors (3) with a 1.5 L working volume. Base medium composition was similar to the previously described base medium but the final nitrogen concentration was defined by combinations of glutamic acid (20, 30, or 40 mM concentrations) and

Table 1: Synthetic medium used for the nutrient requirement studies of
Clostridium acetobutylicum ATCC#824

Compound	Concentration g/100 mL
KH_2PO_4	0.200
$MgSO_4$	0.040
$MnSO_4$	0.001
$FeSO_4$	0.001
NaCl	0.001
Biotin	0.000001
p-aminobenzoic acid	0.000100
Glucose	5.000
N-source	various

ammonium chloride (0 ; 2.50; 5.00 or 10 mM concentrations) . Initial pH of the medium after autoclaving was 5.8 to 6.0 .The sterile medium was flushed with N_2 for 24 hours at a flow-rate of 100 ml/min. The fermentation was performed at 37°C without agitation. Samples were taken throughout the fermentation and analysed by the subsequently described methods.

CALCIUM AND PHOSPHATE EXPERIMENTS: The initial calcium and phosphate studies were performed in 50 mL vials filled with 25 mL of medium and incubated at 37°C in an anaerobic jar. Samples were taken after 72 and 144 hours. The base medium was the same as for the nitrogen studies except the concentration of glutamic acid was 30 mM , the concentration of ammonium chloride was 15 mM and potassium was supplied in the form of KOH (0.34g/L). The pH of the medium after autoclaving was 5.5 to 6.0 . Phosphate was supplied as a combination of mono- and di- basic sodium phosphate with concentration ranging from 0 mM to 100 mM of total phosphate.For the calcium - phosphate studies ,the medium was the same as above except phosphate was in concentrations of 10, 30, 50 or 70 mM (combination of both mono- and di-basic phosphates) with 0, 2.5, 5.0, or 20 mM calcium added as calcium chloride. Initial pH after autoclaving was 6.0 .

SAMPLING AND ANALYSIS: Cells were separated from the medium by centrifugation at 1200xg for 20 minutes at 22°C. Supernatant was removed and used to determine pH, and to determine acetone, ethanol, butanol, acetic acid , butyric acid and glucose concentrations. Solvents and acids were determined by using gas chromatography.Glucose concentration was determined by using HPLC technique. Cells were resuspended in distilled water and the absorbance was determined at 650 nm.

RESULTS AND DISCUSSION

NITROGEN STUDIES: Most practically oriented studies using <u>Clostridium acetobutylicum</u> have tried to optimize or maximize solvent production using readily available agricultural or

industrial substrates with undefined nitrogen and carbon sources
(4). Recent published work has utilized ammonium acetate as the
nitrogen source in a defined medium in batch and continuous
laboratory scale fermentations (1,2).Results presented in Table 2
indicate that the use of glutamic acid as the nitrogen source in
a defined medium supported cell growth (A_{650}=0.961) and solvent
production (total solvent concentration= 1.1 g/100 ml) while
ammonium acetate medium, under the same condition, supported the
solvent production (0.99 g/100 ml) but not much growth (A_{650}=
0.429). Medium, with aspartic acid as the nitrogen source,
initiated cell growth and solvent production while media with
ammonium chloride,ammonium sulfate,alanine or lysine at the
described concentrations (30 mM each) did not induce growth or
solvent production. These results indicate that glutamic acid is
a more suitable nitrogen source for *Clostridium acetobutylicum*
than ammonium acetate or the other organic and inorganic nitrogen
sources that were tested. Further data have indicated that

Table 2.:Effect of various nitrogen sources on cell growth and solvent
production by *Clostridium acetobutylicum* ATCC#824 grown in defined
synthetic medium.

Nitrogen source	A_{650}	ButOH	ACE	ETOH	Total solvent	Butyric acid	Solvent ratio (B : A : E)
Ammonium chloride	0.017	0.037	0.038	0.007	0.082	0.061	5.4 : 5.4 : 1
Ammonium sulfate	0.238	0.031	0.035	0.007	0.073	0.059	4.4 : 5.0 : 1
Ammonium acetate	0.429	0.708	0.159	0.118	0.985	0.039	6.0 : 1.3 : 1
Glutamic acid	0.961	0.730	0.218	0.137	1.085	0.037	5.3 : 1.6 : 1
Aspartic acid	0.553	0.453	0.176	0.044	0.673	0.091	9.9 : 4.0 : 1
Alanine	0.020	0.036	0.037	0.008	0.081	---	4.6 : 4.6 : 1
Lysine	0.018	0.035	0.035	0.000	0.076	---	5.8 : 5.8 : 1

solvent production rates could be increased when ammonium
nitrogen was used with glutamic acid as the nitrogen source in a
defined medium. Results presented in Table 3 show that at three
glutamic acid concentrations (20, 30 and 40 mM) the addition of
small amounts of ammonium nitrogen (up to 5 mM ammonium chloride)
shortened the lag phase as well as the total fermentation time
while maintaining solvent production.However, larger
concentrations of ammonium chloride (10 mM) prolonged the
fermentation time. Highest solvent concentrations with the
shortest lag phase were determined at 20 - 30 mM glutamic acid
and 2.5 - 5.0 mM ammonium chloride concentrations. The
fermentation had a 5 hour lag phase and 45 to 50 hours of total
fermentation time. The butanol concentration usually reached 1.0
to 1.2 % (w/v).
Figures 1, 2 and 3 present data from a batch fermentation of
Clostridium acetobutylicum in optimized defined synthetic medium
with glutamic acid (20 mM) and ammonium chloride (5 mM) as
the nitrogen source. The data represent classical changes in
absorbance , solvents and acids production. That is,growth was
initiated and followed by acid production. When production of

171

TABLE 3.: Growth and solvent production by *Clostridium acetobutylicum* #824 in synthetic medium with glutamic acid and ammonium chloride as nitrogen sources.

GLU:NH$_4$$^+$	time hours	OD 650 nm	Butanol g/100 mL	Acetone g/100 mL	Butyric acid g/100 mL	lag-phase hours
20:0.0	96	0.795	1.10	0.46	0.14	10
30:0.0	75	1.035	1.05	0.46	0.16	10
40:0.0	75	1.000	1.04	0.47	0.16	15
20:2.5	70	0.790	1.20	0.53	0.10	14
30:2.5	68	0.760	1.21	0.57	0.10	5
40:2.5	68	0.760	1.21	0.57	0.11	13
20:5.0	48	0.800	1.02	0.49	0.17	5
30:5.0	45	0.930	1.00	0.46	0.16	5
40:5.0	44	0.950	0.96	0.44	0.17	10
20:10.0	93	0.815	1.02	0.375	0.17	25
30:10.0	70	0.880	1.03	0.44	0.16	26
40:10.0	93	0.770	1.14	0.49	0.19	65
Control	96		0.72	0.325	0.11	--

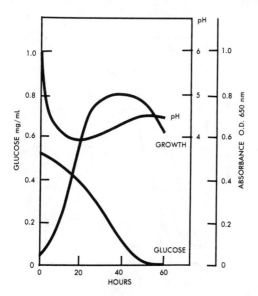

FIGURE 1

BATCH CULTURE OF CLOSTRIDIUM ACETOBUTYLICUM
ATCC#824 - 2L FERMENTOR

172

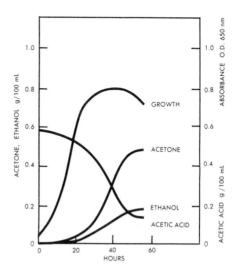

FIGURE 2

BATCH CULTURE OF CLOSTRIDIUM ACETOBUTYLICUM
ATCC#824 - 2L FERMENTOR

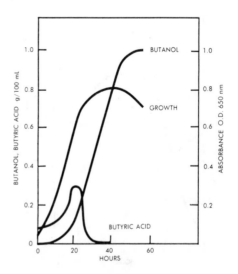

FIGURE 3

BATCH CULTURE OF CLOSTRIDIUM ACETOBUTYLICUM
ATCC#824 - 2L FERMENTOR

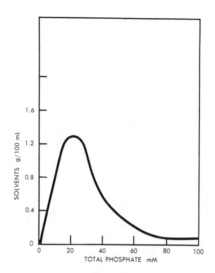

FIGURE 4

BATCH CULTURE OF CLOSTRIDIUM ACETOBUTYLICUM
ATCC#824
EFFECT OF PHOSPHATE ON SOLVENT BIOSYNTHESIS
SAMPLING AT 144 HOURS

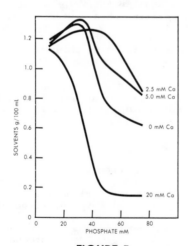

FIGURE 5

BATCH CULTURE OF CLOSTRIDIUM ACETOBUTYLICUM
ATCC#824
EFFECT OF PHOSPHATE-CALCIUM INTERACTION ON
SOLVENT BIOSYNTHESIS

solvents increased the concentration of acids declined. Maximum solvent production under the described condition was obtained between 45 and 48 hours.

CALCIUM - PHOSPHATE STUDIES: Calcium and phosphate seem to be physiological requirements of *Clostridium acetobutylicum* . The physico-chemical characteristics of the interaction of calcium and phosphate are known , therefore studies were attempted to define their interaction in physiological conditions and their effect on metabolite production by *Clostridium acetobutylicum* ATCC#824 . Fig.4 shows that solvent production increased as the phosphate concentration increased. Maximum solvent production was reached with 10 to 30 mM of phosphate in medium when calcium was omitted from the medium. However, higher phosphate concentrations reduced the solvent production.

The calcium-phosphate interaction, when related to the solvent production, showed that low concentrations of calcium (2.5 and 5.0 mM) have little or no effect when in the presence of 10 to 30 mM phosphate. Media with 2.5 and 5.0 mM of calcium produced higher solvent concentrations at phosphate concentrations of 30 mM , 50 mM and 70 mM (Fig.5) . 20 mM concentrations of calcium produced an inhibitory effect on solvent production in media with phosphate concentrations higher than 10 mM.

CONCLUSIONS

1. Glutamic acid can be used as nitrogen source for the acetone-butanol fermentation by *Clostridium acetobutylicum* ATCC#824.

2. Low concentrations of ammonium chloride (2.5 to 5.0 mM) reduced the lag phase and the total fermentation time in batch acetone-butanol fermentation when glutamic acid was used as the major nitrogen source.

3. The required concentrations of phosphate for optimum solvent production were detected in the range of 10 to 30 mM.

4. Interaction of calcium (2.5 to 5.0 mM concentrations) with phosphate (50 to 70 mM concentrations) in the synthetic medium resulted in an improved solvent production in batch acetone-butanol fermentations.

REFERENCES

1. Monot, F. ; J-R. Martin; H.Petitdemange and R. Gay: (1982), Appl.Environ.Microbiol., 44:1318-1324.

2. Monot, F.; J.M. Engasser: (1983), Eur. J. Appl. Microbiol. Biotechnol.,18(4):246-248.

3. Veliky I.A.;S.M.Martin: (1979), Can.J.Microbiol.,16:223-226.

4. Spivey, M.J.: (1978), Process Biochem., 13(2-4):25.

The Use of Renewable Carbohydrate Sources for the Microbial Production of Acetone and Butanol

C.W. Forsberg[1], S.F. Lee[1], H. Schellhorn[1], L.N. Gibbins[1], E. Mason[2], and F. Maine[2]. Department of Microbiology[1], University of Guelph, Guelph, Ontario, N1G 2W1, and Frank Maine Consulting[2], 71 Sherwood Drive Guelph, Ontario, N1E 6E6

ABSTRACT

The optimum conditions for the release of reducing carbohydrate from horticultural grade peat by the Stake process (Stake Technology Inc., Oakville, Ontario) was found to be 3 mins exposure to 200°C prior to explosive decompression. The resulting extract contained galactose, glucose, xylose, mannose, arabinose, and cellobiose, together with a mixture of (presumed) oligosaccharides. The sugars and oligosaccharides accounted for 76% of the dry weight of the material released, while the sugars accounted for 26% of the dry weight of the solubilised material. Two metabolically versatile strains of Clostridium butylicum (NRC 33007 and NRRL B593) fermented up to 30% of this carbohydrate mixture. The products were predominantly acetate and butyrate, with lesser amounts of acetone and butanol. When C. acetobutylicum ATCC 824 was grown in a chemostat with glucose as the growth limiting nutrient at pH 6.5, acetate and butyrate were the major products, while at pH 4.5 the major products were acetone, butanol and ethanol. At this low pH, 73% of the glucose carbon was recovered in solvents. The yield of solvents was lower with other sugars. Strains of Clostridium spp. having polysaccharide hydrolytic activity were sought. C.acetobutylicum ATCC 824 and NRRL B527 produced extracellular endoglucanase and xylanase activities, but were unable to grow with either acid swollen cellulose or xylan as the source of carbohydrate in batch culture. Mutants of ATCC 824 grew on xylan and utilized 70% of the xylan carbohydrate during growth. They had developed a nutritional requirement for bicarbonate. The study of these physiological properties is continuing.

INTRODUCTION

The microbial synthesis of chemicals and fuels is expected to be of major economical significance as supplies of fossil fuels continue to be depleted. A major untapped source of feed stock carbohydrate for such synthesis is peat (Fuchsman, 1980), a natural, and renewable, resource with which Canada is highly endowed. In fact, it is estimated that Canada may have between 11 to 36 % of the total world peat resources (Ismail, 1982).

Peat is a very complex mixture of materials the composition of which is largely dependent upon the degree of degradation due to ageing. Young, undecomposed peats near the surface are relatively rich in carbohydrate, and may contain up to 20% or more of non-cellulose carbohydrate, and a similar amount of cellulosic material. Peat harvested from the lower, and older, layers is highly decomposed, and usually practically devoid of carbohydrate (Fuchsman, 1980). The principle chemical characteristics of peat are shown in Table 1, while sugars commonly detected in peat are listed in Table 2. Although the composition of the peat will vary depending upon its source and extent of humification it is noteworthy that a wide range of sugars is present.

Peat has been tested in several laboratories as a substrate for the acetone-butanol fermentation. It has been used in various forms; (a) without extensive treatment (Fogarty and Ward, 1970); (b) after heating in aqueous suspension at up to 141°C (Kust et al., 1976); and (c) after hydrolysis in 1.5 N H_2SO_4 at 140°C (Le Duy et al., 1983). Le Duy et al. (1983) found that, when an acid hydrolysate containing peat sugars at a concentration of 3 to 5% was used as a substrate for the growth of C. acetobutylicum, practically no solvents were detectable as end-products. This work was not continued (Le Duy, personal communication).

In the present study, steam explosion was used to release non-cellulosic carbohydrates from peat, and this material was then used as the substrate for the acetone-butanol fermentation. To resolve some of the problems associated with the use of such a complex substrate in this fermentation, a program of study including (i) the manipulation of culture conditions, and (ii) strain development and selection, has been undertaken.

MATERIALS AND METHODS

Horticultural peat (Acadia Peat Moss Ltd., Lameque, New Brunswick), that had been sieved through a 4 mm mesh screen, was used as the source of substrate.

The cultures, media, and enzymatic assay methods used were previously described by Lee et al. (1984). Sugars were determined using a Waters HPLC system (Water Scientific Ltd., Toronto), fitted with a Biorad HPX-87P "wood hydrolysate" column (Bio-rad Laboratories Canada Ltd., Mississauga, Ontario), and a high sensitivity ERMA refractive index detector (Erma Optical Works Ltd., Tokyo, Japan).

The "steam-explosion" treatments of samples were conducted by Stake Technology Ltd., Oakville, Ontario.

Table 1: Composition of High-Moor Sphagnum Peats and Low-Moor Sedge Peats (% Dry Weight of Peat).

Component	Sphagnum high-moor peats	Sedge low-moor peats
Ash	1.5- 3.0	7.7-14.5
Bitumen (benzene-ether extract)	3.1- 9.1	3.2- 3.9
Pectins (hot water extract)	4.2- 7.8	2.5- 3.6
Polyoses (hydrolysate with 2% HCl)	9.8-40.4	12.3-21.0
Reducing sugars in polyoses (x 0.9)	8.9-22.5	8.6-11.6
Cellulose (hydrolysate with 80% H_2SO_4)	10.3-23.7	7.8- 8.1
Reducing sugars in cellulose (x 0.9)	9.1-22.4	4.7- 6.4
Residue (lignins and humic acids)	26.3-64.3	56.1-62.2
Total reducing sugar, calculated as glucose	20.0-49.9	16.3-20.0
Protein (N x 6.25)	5.6- 6.9	10.0-13.8

From C.H. Fuchsman p. 80, 1980.

Table 2: Monosaccharide Composition of Peat Extracts (% of total monosaccharides).

	H_2SO_4		HCl	
	(a)	(b)	(c)	(d)
Glucose	39.8	33.5	19.9	18.6
Galactose	15.4	21.8	30.8	28.5
Rhamnose	18.6	11.9	3.1	6.5
Mannose	10.8	6.5	0.3	9.8
Arabinose	-	7.7		
Xylose	15.5	22.1	36.0	36.6
Total Hexoses	84.6	60.0	53.8	53.6
Total Pentoses	15.4	40.0		

(a) Zommers et al. (1974).
(b) Bogdanovskaya et al. (1973).
(c) Bystraya and Rakovskii (1968).
(d) Bystraya and Rakovskii (1968).

From A. LeDuy, 1981. Proc. Int. Peat Symp., Bemidji.

RESULTS

Treatment of Peat

Without pretreatment, peat is an unsatisfactory source of carbohydrate for growth and solvent production by Clostridium acetobutylicum (Fogarty and Ward, 1970). Steam explosion was tested as a method for destabilization of the structure of peat and solubilization of carbohydrate, since it has been successfully used on wood chips for this purpose (Saddler et al., 1982). From the data in Table 3, it can be seen that treatment at $200^{\circ}C$ for at least one minute was essential for the release into solution of a substantial proportion of the dry weight of peat. The total carbohydrate released accounted for approximately 75% of the materials solubilized. Reducing sugar accounted for approximately half of the carbohydrate released, suggesting that the chain lengths of the oligosaccharides released were short. Steam explosion at $215^{\circ}C$ did not enhance the release of carbohydrate. To ensure uniformity of treatment, it was decided to perform all subsequent steam explosions at $200^{\circ}C$ for 3 min, since this maximized the release of reducing sugar (Table 3).

Analysis, by HPLC, of the neutral sugars released by the steam explosion process demonstrated that galactose was the major sugar with lesser quantities of glucose, xylose, mannose, arabinose and cellobiose (Table 4). The sugars accounted for 56% of the total reducing sugar or 26% of the dry weight solubilized. There were also several unidentified peaks which probably were oligosaccharides. After acid hydrolysis, uronic acids and glucosamine were also detected in the solubilized material.

Because of the wide range of sugars available for growth of C. acetobutylicum strains, a large number of cultures was tested for the ability to use various neutral sugars, uronic acids, and amino sugars. Two of the metabolically more versatile cultures, namely, C. butylicum NRC 33007 and C. butylicum NRRL B593, were selected for further study. Initially, cultures were tested for growth and carbohydrate utilization when grown with peat extract as the source of carbohydrate in the medium. After serial subculture at lower concentrations of peat extract, these bacteria grew at concentrations sufficient to give a final carbohydrate concentration of two percent. Higher concentrations were not tested because of a shortage of peat extract. To determine the maximum amount of peat extract carbohydrate that the bacteria could use as a substrate, they were grown with peat carbohydrate as the growth-limiting nutrient. Under these conditions, 20 to 23% of the total carbohydrate was fermented (Table 5). The reducing sugar was used to a greater extent than the total carbohydrate. Galactose was the major sugar remaining after growth (Table 6). In other experiments in which peat extract carbohydrate was included in the culture medium at 1.0% (wt/vol), up to 30% of the carbohydrate was fermented by C. butylicum B593. The major products were acetate and butyrate, with small amounts of acetone and butanol (Fig. 1). Again, insufficient peat extract was available to permit optimization of the fermentation.

Table 3: Influence of treatment time on the release of
soluble material, soluble carbohydrate, and
reducing sugar from peat during the steam
explosion process.

Treatment		pH	Materials Solubilized (%)[1]		
°C	min		Dry wt.	Carbohydrate	Reducing Sugar
Untreated (control)		4.9	0.6	0.4	0.1
200	0.5	3.4	13.8	9.2	3.8
200	1.0	3.3	23.1	17.3	7.1
200	2.0	3.4	23.8	18.1	9.7
200	3.0	3.4	23.9	18.1	10.9
200	5.0	3.4	25.8	18.8	9.4

[1] Each value represents the quantity released from a
15 g sample (50% w/v mositure content).

Table 4: Sugars released from peat by steam explosion at
200°C for 3 min.

Sugar	% of Peat dry wt.	% of total monosaccharides
Cellobiose	0.4	6.9
Glucose	1.0	16.4
Xylose	1.0	16.1
Galactose	2.5	41.6
Arabinose	0.5	8.0
Mannose	0.7	11.0
TOTAL	6.1	100.0

180

Table 5: Carbohydrate and reducing sugar used by *Clostridium butylicum* B593 and NRC 33007 during growth in the chemically defined medium described by Lee *et al*. (1984) with peat extract as the source of carbohydrate.[1]

Strain	Carbohydrate concn in medium (%)	Carbohydrate used (%)	Reducing Sugar used (%)
B593	0.3	20	39
	0.6	20	36
33007	0.3	23	51
	0.6	21	42

[1] Cultures were incubated for 120 h at 35°C.

Table 6: Monosaccharides from peat extract remaining after growth of *C. acetobutylicum* 33007 as described in Table 5.

Time	Monosaccharide	% Remaining
72	Cellobiose	23
	Glucose	5
	Xylose	0
	Galactose	88
	Arabinose/Mannose	50
120	Cellobiose	23
	Glucose	0
	Xylose	0
	Galactose	57
	Arabinose/Mannose	22

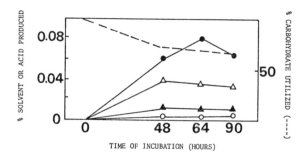

TIME OF INCUBATION (HOURS)

Figure 1: Carbohydrate utilization and solvent production by <u>Clostridium</u> <u>butylicum</u> NRRL B593 during growth in the chemically defined medium described by Lee <u>et al.</u> (1984) containing 1.0 % (wt/vol) peat extract carbohydrate as the sole source of carbohydrate. The initial pH was 6.5. Butyric (●), acetic (△), butanol (▲) and acetone (○).

Improving Carbohydrate Utilization

When <u>C. acetobutylicum</u> 824 was grown in continuous culture on a chemically defined medium with glucose as the limiting nutrient, the pattern of fermentation products was determined by the culture pH (Table 7). At high pH, acetate and butyrate were the predominant fermentation products. However, at pH 4.5, acetone, ethanol, and butanol were the predominant products, and they accounted for 73% of the carbon recovered. With xylose as the source of carbohydrate during culture at pH 4.5, a high conversion of substrate to product was observed, although solvent production was low. In the case of arabinose, both carbon recovery and solvent production was low at pH 4.5.

Strain Selection and Development for Carbohydrate Utilization

In view of the complex array of carbohydrates, including monosaccharides and oligosaccharides, in peat extract, it is essential to ensure that the bacteria used for fermentation studies have extracellular hydrolases to degrade polymers, and that the resulting sugars, made available from all such sources, will be readily fermented.

The efficient use of these materials could be effected through the use of strains of micro-organisms having cellulolytic and hemicellulolytic activities. Several strains of <u>C. acetobutylicum</u> and <u>C. butylicum</u> have been screened for both cellulase (Lee <u>et al.</u> 1984) and xylanase activity. No strains having the ability to hydrolyse and grow on crystalline cellulose were detected, although two which produced extracellular endoglucanase activities capable of hydrolysing amorphous cellulose, were identified (Table 8). These strains, <u>C. acetobutylicum</u> ATCC 824 and NRRL B527, were unable to grow on amorphous cellulose, for some unknown reason. The strain B527 grew on the products arising from the action of its own endoglucanase on amorphous cellulose. Both strains produced endoglucanase during carbon-limited growth with glucose, cellobiose, xylose, or mannose as the substrate. Thus, it would be expected that the presence of endoglucanase would contribute to the hydrolysis of cello-oligosaccharides in peat extract.

Table 7:   Solvent production by C. acetobutylicum 824 during chemostat culture with a carbohydrate as the limiting nutrient and a dilution rate of 0.05[1].

Carbohydrate source	pH	Acetone	Ethanol	Butanol	% Conversion to Solvents	% Carbon[2] recovery
Glucose	5.2	0.02	3.7	6.2	9.3	65.4
	4.5	7.0	6.3	57.2	73.0	82.8
Xylose	5.2	0.5	2.2	1.2	4.3	54.6
	4.5	11.5	2.7	15.2	23.4	67.9
Mannose	5.2	1.6	2.8	3.6	6.0	53.6

[1] The concentrations of all products are expressed as millimolarity.
[2] The percentage of carbohydrate carbon recovered as fatty acids and solvents.

Table 8:   Glucanase and glycosidase activities of C. acetobutylicum ATCC 824 grown on xylose in a chemostat at pH 5.2 and a dilution rate of 0.05.

	Extracellular	Cell-bound
Endoglucanase[1]	121.7 (485)	26.2 (43)
Cellobiase	8.1 (32)	33.6 (56)
Xylanase	295.9 (1179)	73.3 (122)
Xylopyranosidase	41.7 (166)	41.9 (70)
∝-L-arabinofuranosidase	25.5 (102)	15.0 (25)
β-D-fucopyranosidase	19.6 (78)	2.7 (5)
β-D-mannopyranosidase	7.0 (28)	9.3 (15)
∝-D-mannopyranosidase	0	0
β-D-glucopyranosidase	27.8 (111)	89.1 (148)
∝-D-glucopyranosidase	0	0
β-D-galactopyranosidase	0	0
∝-D-galactopyranosidase	0	0
Polygalacturonase	55.4 (221)	2.2 (4)
Mannanase	0	0

[1] Glucanase and xylanase activities expressed as nmol reducing sugar per min per ml. Glycosidase activities are expressed as nmol of product per min per ml. Values in parenthesis are expressed as specific activity (nmol per min per mg protein).

From Lee et al. 1985.

Hemicellulolytic activity (expressed as xylanase activity) is widely distributed among the solvent-producing clostridia. The C. acetobutylicum strain ATCC 824 was shown to possess xylanase, xylopyranosidase, arabinofuranosidase, fucopyranosidase, mannopyranosidase activities, and, in addition, polygalacturonase activity (Table 8). Although the C. acetobutylicum strains ATCC 824 and NRRL B526 hydrolysed xylan, they were unable to use it as the sole source of carbohydrate in batch culture. An attempt was made to obtain, from these strains, mutants capable of growth on amorphous cellulose and xylan. Strain B527 was found to be very refractory to mutagens. Strain ATCC 824 gave rise to rifampicin-resistant mutants, although no mutants able to grow on amorphous cellulose as carbon source were obtained. However, mutants able to grow with xylan as the sole source of carbohydrate were obtained. These derivatives had a requirement for enzymically hydrolyzed casein; this requirement was not relieved by "casamino acids", but was satisfied by bicarbonate. From the data presented in Fig. 2 it can be seen that the amount of growth of mutant XYN 1 increased as the bicarbonate concentration was raised to 0.2%. At 0.4% bicarbonate, growth was reduced. During growth in the bicarbonate medium, up to 70 % of the larchwood xylan carbohydrate was utilized. The mutant grew well at pH 5.5 in the chemostat, with xylan as the limiting nutrient, but growth ceased and washout occurred when the pH was allowed to drop to 5.0. Work is continuing to determine both the basis for the requirement for bicarbonate and the corresponding physiological differences between the parent and the mutant strain. Subsequently it was found that ATCC 824 and NRRL B526 would grow with xylan as the sole source of carbohydrate in chemostat culture with pH control (Lee et al. 1985).

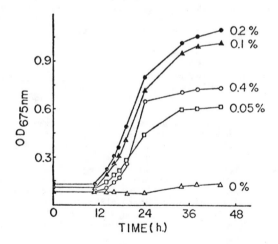

Figure 2: Growth of Clostridium acetobutylicum XYN 1 mutant in the chemically defined medium described by Lee et al. (1984) with 1 % xylan as the source of carbohydrate in the presence of various concentrations of sodium bicarbonate. The bicarbonate was added as neutralized sodium carbonate; the initial pH was 6.5.

DISCUSSION

The prerequisites for an economically successful fermentation of lignocellulosic materials are appropriate pretreatment of the substrate, correct strain selection, and appropriate culture conditions. The steam explosion process, which is known to be an effective pretreatment for woody materials (Saddler et al., 1982), was used as an initial treatment to destabilize the structure of the lignocellulose and to release hemicellulosic materials from horticultural grade peat. Using this method, up to 26% of the peat was converted to a soluble form, the bulk of which (75%) was carbohydrate (also see Forsberg et al. 1986). The high reducing power of the material suggested that the polysaccharide had short chain lengths. Approximately 33% of the total carbohydrate was present as free monosaccharides, consisting of galactose, glucose, xylose, mannose, arabinose, and cellobiose in descending order of concentration.

The solubilized peat materials was found to be non-toxic to the Clostridium strains tested at concentrations providing 2% of carbohydrate. In contrast, water extracts from steam-exploded wheat straw and from aspen wood chips were reported to be highly toxic, thus restricting growth of the organism and the resulting enzymatic hydrolysis of the pretreated substrate and fermentation of the derived sugars (Mes-Hartree and Saddler, 1983).

The clostridia tested were able to use only 20 to 30% of the carbohydrate included in the culture medium. The simplest reason for this low efficiency is that the C. butylicum strains lack many of the enzymes for the degradation of the oligosaccharides in the peat extract. The products of fermentation of the carbohydrates were predominantly acetate and butyrate, with low concentrations of acetone and butanol. This distribution of products presumably reflected an inefficient transformation of fatty acids to solvents during the second phase of the batch culture (Spivey, 1978). This problem probably could be corrected by improved pH control, as has been reported by Bahl et al. (1982a & b; Huang et al. 1986) for enhancing the production of solvents by C. acetobutylicum during continuous culture. Also, given the complex mixture of carbohydrates, catabolite repression by glucose of the pathways of catabolism of other carbohydrates might be expected. However, it was claimed by Ounine et al. (1983) that catabolite repression was not a problem, although a different mixture of substrates was studied in these latter experiments. Continuous culture permits the maintenance of a stable low pH to enhance solvent production, whether phosphate or carbon limitation is used. Carbon-limited growth in the chemostat should decrease the problem of catabolite repression that might result as a consequence of the presence glucose along with other sugars and polysaccharide materials. The great selectivity of the chemostat system (Dykhuizen and Hartl, 1983) should also permit the enrichment of variants of C. acetobutylicum and other organisms capable of fermenting those carbohydrates which are normally only metabolised with difficulty.

The role of bicarbonate in the growth of various C. acetobutylicum and C. butylicum strains is not well understood. Many strains appear to have a requirement for bicarbonate for the initiation of growth in batch culture, but, once growth of the organism has been established, the endogenous production of carbon dioxide apparently meets their continued requirements for this metabolite. The bicarbonate requirement of the mutant XYN 1 for growth on xylan was unexpected since the parent strain grew with xylose as the sole source of carbohydrate. The mechanism of replacement of the bicarbonate requirement by a enzymatically hydrolyzed casein is unknown. This relationship undoubtedly is important since growth of C. acetobutylicum at low pH (4.5) enhances solvent production (Bahl et al., 1982). At this pH, bicarbonate is lost from the medium as carbon dioxide and therefore could not meet substrate requirements.

The conditions necessary for the maximum conversion of sugar to solvents is somewhat different for each sugar, and also the quantitative distribution of products from these sugars can be different.

The chemostat is the preferred culture system for studying solvent production by Clostridium strains, as Bahl et al. (1982a,b) have clearly demonstrated that the maintenance of a stable low pH (4.3 to 4.5) is very critical for maximizing solvent production. The metabolism of glucose in a carbon-limited culture at low pH permitted the achievement of a combination of high conversion to solvent (73%) with a carbon recovery in fermentation products of 83%. In batch culture systems, usually about 30% of the substrate is converted to solvents, while approximately 50% of the carbon is lost as carbon dioxide (Spivey, 1978). With xylose or cellobiose as the carbon source, the conversion of substrate to solvent was low by comparison to that observed from glucose, except that growth of C. acetobutylicum on xylose at low pH produced a greater yield of total products. Thus, in the chemostat, the increased fermentation product yield may perhaps be attributed to more efficient recycling of metabolic carbon dioxide. However, the underlying reason for the differences in solvent yield from the various carbohydrates is not yet apparent. The resolution of this question is an important prerequisite to further improvement of these fermentation systems.

ACKNOWLEDGEMENT

This research was funded by a contract from the National Research Council of Canada to Frank Maine Consulting Ltd., and a Strategic Grant from the Natural Sciences and Engineering Research Council of Canada to C.W.F. and L.N.G.

REFERENCES

Bahl, H., W. Andersch, K. Braun, and G. Gottschalk. 1982a. Effect of pH and butyrate concentration on the production of acetone and butanol by Clostridium acetobutylicum grown in continuous culture. Eur. J. Appl. Microbiol. Biotechnol. 14: 17-20.

Bahl, H., W. Andersch, and G. Gottschalk. 1982b. Continuous production of acetone and butanol by Clostridium acetobutylicum in a two-stage phosphate limited chemostat. Fur. J. Appl. Microbiol. Biotechnol. 15: 201-205.

Bogdanovskaya, Zh.N. 1973. Effect of culture conditions on the growth of feed yeasts and on the synthesis of Carotenoids (Russian). Mikroorganizmy-Produtsenty Biol. Aktiv. Veshchestv, 72-77.

Bystraya, A.V., and V.E. Rakovskii. 1968. Partition paper chromatographic study of carbohydrate composition of peat after thermohydrolysis (Russian), Kompleks. Ispol'z Torfa 2: 32-37.

Dykhuizen, D.E., and D.L. Hartl. 1983. Selection in chemostats. Microbiol. Rev. 47: 150-168.

Fogarty, W.M., and J.A. Ward. 1970. The influence of peat extract on Clostridium acetobutylicum. Plant and Soil 32: 534-537.

Forsberg, C.W., H.E. Schellhorn, L.N. Gibbins, F. Maine and E. Mason. 1986. The release of fermentable carbohydrate from peat by steam explosion and its use in the microbial production of solvents. Biotechnol. Bioeng. 28: 176-184.

Fuchsman, C.H. 1980. Peat Industrial Chemistry and Technology. Academic Press, Toronto, Canada.

Huang, L., C.W. Forsberg, and L.N. Gibbins. 1986. Influence of external pH and fermentation products on the intracellular pH and cellular distribution of fermentation products. Appl. Environ. Microbiol. In Press, to appear in the June issue.

Ismail, A. 1982. Canadian peat commercialization. pp.607-626. In Peat as an energy althernative II. Institute of Gas Technology, IIT Center, 3424 South State Street Chicago, Illinois 60616.

Kuster, E., J. Rogers, and A. McLoughlin. 1968. Stimulating effect of peat extracts on microbial metabolic reactions pp. 23-27. Proc. Third Int. Peat Cong. Quebec.

Le Duy, A., J.M. Boa, and A. Laroche. 1983. SCP, polysaccharide, and alcohol from peat hydrolyzate. Presented at National Spring Meeting of the American Institute of Chemical Engineers, Houston, Texas, U.S.A., March 27-31, 1983.

Lee, S.F., C.W. Forsberg, and L.N. Gibbins. 1984. Carboxymethylcellulase and cellobiase activities of Clostridium acetobutylicum and C. butylicum strains. Fifth Canadian Bioenergy R & D Seminar, S. Hasnain, ed., pp. 569-572. Elsevier Applied Science Publishers Ltd., Barking, England.

Lee, S.F., C.W. Forsberg, and L.N. Gibbins. 1985. Xylanolytic activity of Clostridium acetobutylicum. Appl. Environ. Microbiol. 50: 1068-1076.

Mes-Hartree, M., and J.N. Saddler. 1983. The nature of inhibitory materials present in pretreated lignocellulosic substrates which inhibit the enzymatic hydrolysis of cellulose. Biotechnol. Lett. 5: 531-536.

Ounine, K., H. Petitdemange, G. Raval, and R. Gay. 1983. Acetone-butanol production from pentoses by Clostridium acetobutylicum. Biotechnol. Lett. 5: 605-610.

Saddler, J.N., H.H. Brownell, L.P. Clermont, and N. Levitin. 1982. Enzymatic hydrolysis of cellulose and various pretreated wood fractions. Biotechnol. Bioeng. 24: 1389-1402.

Spivey, M.J. 1978. The acetone/butanol/ethanol fermentation. Proc. Biochem. 13: 2-5.

Zommers, Z., E. Trusle, and L.M. Iosifova. 1974. Chemical composition of a peat hydrolyzate (Russian), Fermentatsiya, 96-101.

PRODUCTION OF ALCOHOL BY FLOCCULATING YEAST

N. Kosaric, A. Wieczorek* and Z. Duvnjak

Chemical and Biochemical Engineering, Faculty of Engineering Science
The University of Western Ontario, London, Ontario
N6A 5B9 Canada

*Institute of Chemical Engineering, Technical University of Lodz
90-924 Lodz, ul Wolczanska 175, Poland

ABSTRACT

A flocculant yeast *Saccharomyces diastaticus* was used for ethanol production
n batch,semi-continuous and continuous fermentation processes.
Jerusalem artichoke and fodder beet were used as a raw material for alcohol
production at 33°C.
The high biomass concentrations in reactors were achieved both in semi-continuous (up to 45g•L^{-1}) and continuous processes (up to 70 g•L^{-1}). As a result the
volumetric ethanol productivity in a semi-continuous process was 3.5 times higher in
comparison to the batch process. The semi-continuous fermentation with a removal of
30% of beer and its replacement with the same amount of fresh medium was completed
in 4 hours.
A very high volumetric ethanol productivity (above 30 g•L^{-1}•h^{-1}) was achieved
in a continuous process with the external cell recycle using Jerusalem artichoke
juice. The productivity was even higher (about 44g L^{-1}•h^{-1}) in a system with internal yeast settling when the dilution rate was 1.162 h^{-1} reaching a yeast concentration of 50-70g•L^{-1}. The ethanol yields ($Y_{P/S}$) of about 0.49 g•g^{-1} were almost
constant at all dilution rates.
Similar results were obtained with fodder beet juice as a substrate.

INTRODUCTION

Ethanol production by fermentation can be operated generally in a batch or a
continuous mode. Industry uses at the present time mainly batch technology. This
technology is simple; however, the reactor productivity is relatively low. An increase in the reactor productivity has been achieved by application of cell recycle.
Continuous systems offer an increase in the reactor productivity which reaches
a level of around 3 times higher in comparison with a batch process.
A number of modifications of continuous systems have been done in order to
ameliorate the volumetric ethanol productivity (Kosaric et al.1983). More productive systems require a higher cells concentration, which has to be realized either
by returning (recycling) separated cells from the effluent to the reactor. With an
effective recycling system, the cell concentration can rise to 100-150 g L^{-1} and the
volumetric ethanol productivities as high as 50 g•L^{-1}•h^{-1} could be attained
(Bu'Lock, 1979).
The simplest systems for cell retention are tower reactors using strongly
flocculant yeasts (Hough and Button, 1972). Their weakness is that if flocculation
fails the whole system becomes uncontrolled. In addition, a significant portion of
the biomass is not utilized as a result of poor mixing. However, with improvements,
this system offers a rather simple way for a productivity increase.

A few other methods of continuous systems exist such as dialysis fermentors (Pirt, 1975), vacuum fermentation (Cysewski and Wilke, 1977), etc, but they have yet to be proven viable on a large scale.

In this study, batch and semicontinuous processes and continuous process with external and internal cell settling systems have been used for the production of fuel ethanol from Jerusalem artichoke and fodder beet juices by a flocculating yeast *Saccharomyces diastaticus*.

MATERIAL AND METHODS

The yeast *Saccharomyces diastaticus* 62 was obtained from Labatt's Brewing Co., London, Ontario, Canada.

The culture was maintained on agar slants containing: malt extract (3 g), yeast extract (3 g), peptone (5 g), glucose (10 g), agar (20 g) and distilled water (to 1 L). After growth the culture was kept at 4oC.

The above medium (from which agar was omitted) was used for preparation of inocula. The inocula were prepared by growing the yeasts for 20 hours at 32oC in 500 ml Erlemmeyer flasks in a rotary shaker. The initial pH of the media was 5.6.

Sterilized J. artichoke juice or fodder beet juice without the addition of supplemental nutrients were used as media for ethanol production.

Ethanol production was carried out in Bellco reactors (1 L) and in a Chemap reactor (14 L).

Carbohydrates were determined spectrophotometrically by a modified anthrone method (Weiner, 1978). Ethanol was determined enzymatically using alcohol dehydrogenase according to Bert and Gutmann (1974). For biomass determination, the culture broth was centrifuged and the biomass washed with water, centrifuged, dried overnight at 105oC and weighed.

RESULTS AND DISCUSSION

In this work the production of ethanol was carried out by a flocculant yeast *Saccharomyces diastaticus*. *S. diastaticus* frequently appears in alcohol broths where its presence is undesireable. It readily produces ethanol as well as certain higher alcohols and esters that impart an off-taste to the alcohol, making it unsuitable as a beverage (Masschelein, 1973). That is the reason why *S. diastaticus* is considered to be a contaminant in beverage alcohol production. The presence of the abovementioned by-products in ethanol is not a problem when ethanol is utilized as a fuel. Consequently, there is no objection for the utilization of this yeast in fuel alcohol production. In addition to the good ethanol production capability, this yeast has a remarkable flocculating ability (Andrews and Gilliland, 1952) that can be utilized in cheap and simple ethanol production systems with high biomass concentrations. Comparing the sensitivity of *S. diastaticus* 62 and that of *S. cerevisiae* 125 to ethanol inhibition, it was found that the former yeast is less sensitive. (Duvnjak and Kosaric, 1981).

Tubers of Jerusalem artichoke and roots of fodder beet were used as a raw material for the fuel ethanol production. J. artichoke can be grown even in a poor soil giving a relatively high yield. In a richer soil and under ideal growing conditions, the yield can be above 70 tons of tubers per hectare (Chubey and Dorrell, 1982). Chemical composition of J. artichoke tubers is shown in Table 1.

Carbohydrates in J. artichokes are fructofuranose units, mainly in the form of inulin. Inulin as it occurs in J. artichokes consists of linear chains of D-fructose molecules united by $\beta(2 \rightarrow 1)$ linkages and terminated by a D-glucose molecule linked to fructose by an $\alpha(1 \rightarrow 2)$ bond. The remainder of the sugars resemble inulin, being composed of essentially fructofuranose units (Bacon and Edelman, 1951).

Carbohydrates of J. artichoke tubers represent about 80% of tubers dry weight and after being hydrolysed are very good C-source for ethanol production by *S. diastaticus*.

Fodder beet is a high-yielding forage crop. It is similar to sugar beet with respect to agricultural requirements. The interest in this crop lies in its higher yield of fermentable sugars per hectare relative to sugar beet and its comparatively high resistance to loss of fermentable sugars during storage. The yield of fodder beet is about 50-150 tons/hectare.

The approximate composition of fodder beet is shown in Table 2 (Kosaric et al., 1984).

Sugars in fodder beet are mainly in the form of sucrose.

Fodder beet roots and Jerusalem artichoke tubers were submitted to an extraction procedure (Figure 1) in order to obtain juices which were used as media for ethanol production.

Table 1. Chemical Composition of J. Artichoke
Tubers (Kaldy et al. 1980)

Dry matter (% of fresh weight)	24.12
Total protein (% of dry weight)	6.53
Total fat "	7.33
Total sugar "	51.65
Starch "	26.26
Fiber "	5.28
P "	0.12
K "	1.44
Ca "	0.39
Mg "	0.11
Fe "	0.0033
Vitamin A (I.U. in fresh weight)	37
Vitamin C (mg/100 g in fresh weight)	0.82

Table 2. Approximate Composition of Fodder Beet

	% Fresh Weight	% Dry Weight
Dry Matter	20	
Moisture	80	
Crude Protein	1.4	7.0
Fat	0.1	0.5
Cellulose	1.2	6.0
Ash	1.1	5.5
Nitrogen Free Extract including:	16.2	81.0
sugar	11.0	55.0

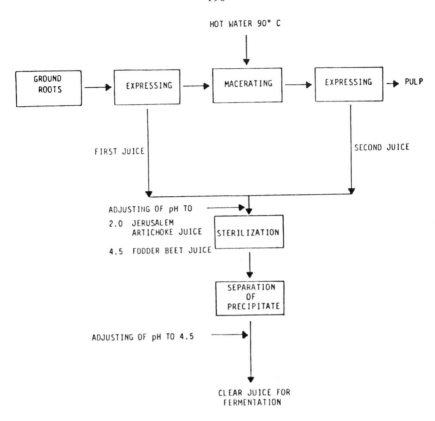

Figure 1. Extraction of Fodder Beet and Jerusalem Artichoke Juice

Production of fuel ethanol from these substrates by *S. diastaticus* 62 was carried out in batch, semi-continuous and continuous fermentation processes.

Conversion of carbohydrates from J. artichoke juice hydrolyzed with sulphuric acid to ethanol in a batch process was carried out in a 1 L Bellco reactor. The medium was inoculated with 10% of biomass. Data for this process are shown in Figure 2.

In the process a maximum concentration of biomass was achieved in about 13 hours and of ethanol in about 20 hours. Ethanol yield was 95.6% of a theoretical value. In a process carried out with the same juice but which was enzymatically hydrolyzed, biomass concentration was about the same as in the acid hydrolyzed juice but ethanol yield was only 87.2%. This can be expected because a high pressure liquid chromatography analysis showed that higher levels of non-fermented sugars (in the form of oligosaccharides), were present in the fermented broth of enzymatically hydrolyzed juice than in that of acid hydrolyzed one (Figure 3).

Figure 4 shows the fermentation of fodder beet juice in a batch process by *S. diastaticus* 62 in a 14 L Chemap reactor.

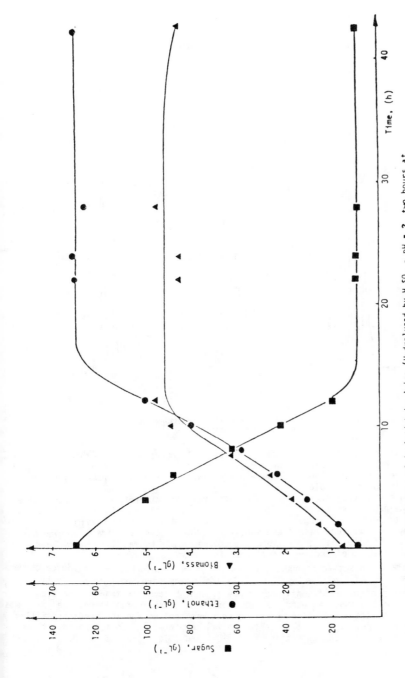

Figure 2. Fermentation of J. Artichoke Juice (Hydrolyzed by H_2SO_4 - pH = 2, two hours at 100ºC) by *S. diastaticus* 62 at 33ºC

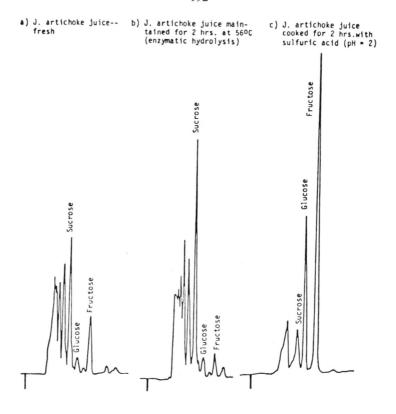

a) J. artichoke juice--
   fresh

b) J. artichoke juice maintained for 2 hrs. at 56°C
   (enzymatic hydrolysis)

c) J. artichoke juice
   cooked for 2 hrs.with
   sulfuric acid (pH = 2)

Figure 3. Qualitative Determination of Carbohydrate Content
in the J. Artichoke Juice

This fermentation was accomplished after 17 hours with the ethanol yield of 92.9% of theoretical value. The overall productivity of the process was 3.9 g of ethanol per litre per hour. Summary of fermentation kinetic parameters for this process is shown in Table 3.

In a semi-continuous process, a batch reactor was operated in batch mode until the process completed. After that agitation was stopped and the biomass was allowed to settle. A large amount of clear beer (about 80% of total) was drawn and replaced with fresh substrate. The process was again carried out batchwise and after its completion, the procedure was repeated. Figure 5 shows a schematic diagram of the semi-continuous fermentation system.

Data for a semi-continuous process of conversion of fodder beet juice carbohydrates by *S. diastaticus* 62 in a 1 L Bellco reactor are shown in Figure 6. For start-up of the semi-continuous process, a high concentration of inoculum (18 g·L^{-1} of biomass) was used. After 6 hours of batch fermentation, the yeast was allowed to settle. A fraction of clear beer was removed and the reactor refilled with fresh juice. The first fermentation was completed in about 5 hours with an ethanol and biomass concentration of 74 g·L^{-1} and 33 g·L^{-1} respectively.

Table 3. Fermentation Kinetic Parameters for
S. diastaticus 62 Using Fodder Beet Juice

Parameter	*S. diastaticus* 62
Initial concentration of sugars, S_0 (% w/v)	13.3
Maximum ethanol concentration, P_{max} (% w/v)	6.63
Biomass concentration in stationary growth phase, X_{max} (g biomass dry wt./L)	8.99
Maximum specific growth rate, μ_{max} (h^{-1})	0.420
Doubling time, T_d (h)	1.65
Biomass yield coefficient from sugars, $Y_{X/S}$ (g biomass dry wt./g sugars utilized)	0.070
Ethanol yield coefficient from sugars, $Y_{P/S}$ (g ethanol/g sugars utilized)	0.50
Ethanol productivity from biomass, $Y_{P/X}$ (g ethanol/g biomass dry wt.)	7.1
Time to complete fermentation (h)	17
Conversion yield, Y (% of theoretical)	92.9
Productivity of fermentation, g ethanol $L^{-1}h^{-1}$	3.9

In following repetitions of the process, the biomass concentration was about 5 g·L^{-1} and only 4 hours was required for fermentation.

The average ethanol yield from sugar utilized was $Y_{P/S}$ = 0.468 g·g^{-1} and the volumetric ethanol productivity was Q_p = 13.5 g·L^{-1}·h^{-1} (based on 4 hour fermentation). This productivity is about 3.5 times that obtained in a simple batch fermentation.

However, considering that the useful volume of the fermentor is only 73% of the total (27% is occupied by the yeasts), the productivity is 9.8 g·L^{-1}·h^{-1}, i.e. 2.5 times that of batch process.

The viability of the cells after ten runs of fermentation was 98% and no contamination occurred.

This semi-continuous process showed the following advantages over the batch:
- high biomass concentration and high productivity
- reduction in fermentation time (to 3-4 hours) in comparison to the conventional (more than 12 hours) in batch operation
- cell separation without centrifugation
- reduction in equipment size due to high productivity of the system
- reduction of contamination by bacteria and its better control
- simple operation with no sophisticated equipment
- favorable energetics and economics.

Continuous process for the fuel ethanol production by *S. diastaticus* 62 was carried out with both external cell recycle and internal settling of cells.

A process with external cell recycle has been accomplished in a two stage

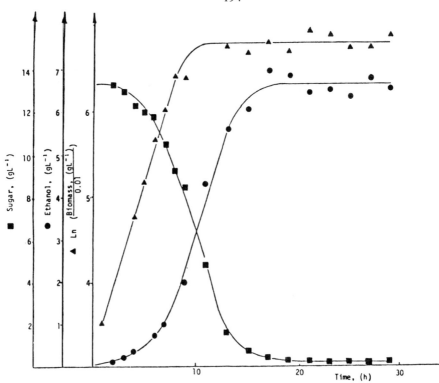

Figure 4.   Batch Fermentation of Fodder Beet Juice by *S. diastaticus* 62

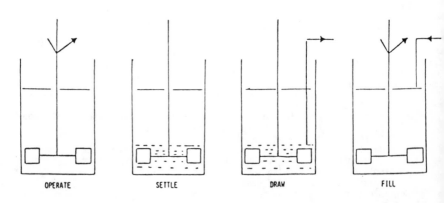

Figure 5.   Schematic Diagram of Semi-Continuous (Fill and Draw) Fermentation System

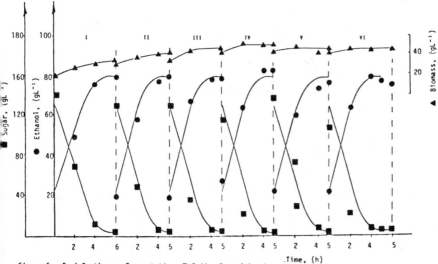

Figure 6. Semi-Continuous Fermentation of Fodder Beet Juice by
Saccharomyces diastaticus 62 in a 1L Belles Reactor

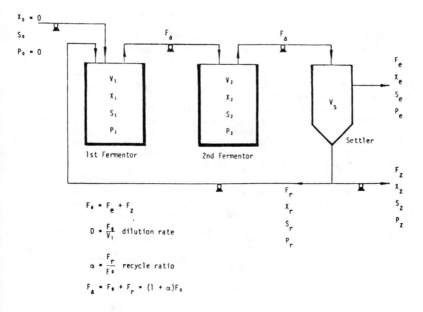

$F_0 = F_e + F_z$

$D = \dfrac{F_0}{V_1}$ dilution rate

$\alpha = \dfrac{F_r}{F_0}$ recycle ratio

$F_a = F_0 + F_r = (1 + \alpha)F_0$

Figure 7. Schematic Diagram of the Fermentation System with "External" Cell Recycle

system composed of two 1 L Bellco reactors and one settler. Figure 7 shows a simplified version of this system.

J. artichoke juice was used in this process. Process parameters for the external cell recycle are shown in Table 4.

Table 4. Fermentation Parameters for External Settling System (**J.** Artichoke Juice Fermented by *S. diastaticus* 62)

F_0 ml h⁻¹	α -	D h⁻¹	S_{T_0} gL⁻¹	S_{T_1} gL⁻¹	S_{T_2} gL⁻¹	P_1 gL⁻¹	P_2 gL⁻¹	Q_{P_1} gL⁻¹h⁻¹	Q_{P_2} gL⁻¹h⁻¹	$(Y_{P/S})_1$ gg⁻¹	$(Y_{P/S})_2$ gg⁻¹
172.0	0.744	0.257	154.4	52.7	40.5	50.2	56.0	12.90	1.49	0.495	0.475
172.0	0.547	0.257	154.4	50.9	35.7	52.4	60.1	13.47	1.99	0.506	~0.510
172.0	0.70	0.20	154.4	43.0	32.0	56.0	61.6	11.2	1.12	0.50	~0.51
486.0	0.90	0.57	171.0	65.4	50.0	52.8	60.5	30.10	4.38	0.50	~0.50

The biomass concentrations in the fermentors were $X_1 = 18.86$ gL⁻¹

$$X_2 = 21.21 \text{ gL}^{-1}$$

The data from Table 4 show that process with external yeast recycle has a very high volumetric ethanol productivity compared to batch and continuous without recycle systems. When dilution rate was D = 0.57 h⁻¹ volumetric productivity of the first reactor was even 30 g·L⁻¹·h⁻¹. The ethanol yields $Y_{P/S}$ were very high (about 0.49 g·g⁻¹. Most of the ethanol was produced in the first reactor; the second one could be even eliminated from the system.

Taking into account difficulties in transfer of concentrated biomass in a system with external recycle, a system with internal settling of yeast cells was developed. The system consists of two reactors (Figure 8) in which settling zones are created with adequate mixing devices and regime and with the incorporation of suitable buffles enabling retention of yeast biomass in the reactors with the upflow of nutrient medium and an overflow of clear beer.

Table 5 shows the parameters of the process.

The concentration of carbohydrate and ethanol and the volumetric ethanol productivity of the system were dependent on the dilution rates. At the dilution rate of D = 1.162 h⁻¹ very high volumetric ethanol productivity (about 44 g·L⁻¹·h⁻¹) was obtained. The ethanol yields $Y_{P/S}$ were also very high (about 0.49 to 0.50 g·g⁻¹) A cell concentration of 50-70 g·L⁻¹ was observed at all dilution rates.

In the first reactor with an increase in dilution rate, the volumetric ethanol productivity increased as well but the percentage of sugar utilized dropped from 78% at D = 0.129 h⁻¹ to 56% at D = 1.162 h⁻¹.

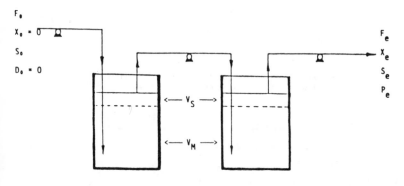

$F_o = F_e$

V_S = volume of settling zone

V_M = volume of mixing zone

$V = V_S + V_M$ = volume of liquid

$D = \dfrac{F_o}{V}$ dilution rate

Figure 8.   Schematic Diagram of the Continuous Process with the Internal Settling System.

Table 5.   Fermentation Parameters for Internal Settling System
(J. Artichoke Juice Fermented by *S.diastaticus* 62)

F_o mL h^{-1}	D h^{-1}	S_o gL^{-1}	First Fermentor				Second Fermentor			
			S_1 gL^{-1}	P_1 gL^{-1}	$Y_{P/S}$ gg^{-1}	Q_p gL^{-1}h^{-1}	S_2 gL^{-1}	P_2 gL^{-1}	$Y_{P/S}$ gg^{-1}	Q_p gL^{-1}h^{-1}
15.5	0.129	161.0	36	61	0.49	7.9	28	65	0.50	0.5
31.0	0.258	161.0	51	54	0.49	13.9	27	66	0.50	3.1
46.5	0.387	161.0	47	56	0.49	21.7	39	60	0.46	1.5
62.0	0.517	161.0	52	55	0.49	28.4	39	60	0.49	2.6
108.5	0.904	164.0	64	49	0.49	44.3	44	59	0.50	9.0
132.5	1.162	164	72	38	0.41	44.1	46	51	0.50	15.1

When fodder beet juice was used as a substrate in the continuous process with internal settling system, similar results were obtained as in the case where J. artichoke juice was used (Table 6).

Table 6. Fermentation Parameters for Internal Settling System
(Fodder Beet Juice Fermented by *S.diastaticus* 62)

| F_o | D | S_o | First Fermentor | | | | Second Fermentor | | | |
| | | | S_1 | P_1 | $Y_{P/S}$ | Q_p | S_2 | P_2 | $Y_{P/S}$ | Q_p |
$mL\ h^{-1}$	h^{-1}	gL^{-1}	gL^{-1}	gL^{-1}	gg^{-1}	$gL^{-1}h^{-1}$	gL^{-1}	gL^{-1}	gg^{-1}	$gL^{-1}h^{-1}$
31.0	0.258	111.8	4.0	53.8	0.50	13.88	2.0	54.0	*	*
62.0	0.517	111.8	5.0	53.8	0.50	27.81	2.0	55.0	*	*
93.0	0.779	111.8	5.0	54.0	0.51	42.07	*	*	*	*
108.5	0.904	111.8	10.0	51.5	0.51	46.56	5.1	54.0	0.51	2.26
124.0	1.033	111.8	14.2	49.5	0.51	51.13	3.2	54.8	0.48	5.48

* not calculated

These results show that the application of continuous systems with internal settling of cells in the ethanol production using a flocculant yeast enables simple retention of biomass in the reactor achieving its concentration of up to 50 - 70 $g \cdot L^{-1}$ even at high dilution rates. Consequently, high volumetric ethanol productivities are realized (up to 50 $g \cdot L^{-1}$).

Some of the advantages of this system with internal settling of cells are:
- high ethanol productivities at high dilution rates
- high ethanol yields
- external yeast recycle and its concentration by centrifugation is avoided
- lower capital investment due to reduction in equipment size and no need for external cell recycling and concentration devices
- simple reactor configuration
- reduction of contamination by bacteria due to high dilution rates.

REFERENCES

Andrews, J. and Gilliland, R.B. (1952). J. Inst. Brewing, 58, 189.

Bacon, J.S.D. and Edelman, J. (1951). The Carbohydrates of the Jerusalem Artichoke and Other Compositae. Biochem. J., 48, 114.

Bert E. and Gutmann, J. (1974). Ethanol Determination With Alcohol Dehydrogenose and NAD. In: Methods in Enzymatic Analysis, H.U. Bergmeyer, ed., Vol. 3, 1499, New York, Academic Press.

Lock, J. (1979). Industrial Alcohol Society for General Microbiology. Symposium, , 309.

ubey, B.B. and Dorrell (1982). Columbia Jerusalem Artichoke. Canadian Journal Plant Science, 62, 537,

sewski, G.R. and Wilke, C.R. (1977). Rapid Ethanol Fermentation Using Vacuum and ell Recycle. Biotechnology and Bioengineering, 19, 1125.

vnjak, Z. and Kosaric, N. (1981). Ethanol Production by Saccharomyces diastaticus. dvances in Biotechnology, M. Moo-Young, ed., Vol. II, 175, Toronto, Pergamon Press.

ough, J.S. and Button, A.M. (1972). Continuous Brewing. Progress in Industrial icrobiology, 11, 89.

aldy, M.S., Johnston, A. and Wilson, D.B. (1980). Nutritive Value of Indian Bread-oot, Squaw-Root and Jerusalem Artichoke. Economic Botany, 34, 352.

osaric, N., Wieczorek, A. and Kliza, S. (1984). Fermentation Ethanol From Agri-ultural Crops. Part I. Experimental Results With Free Cell Systems. Final Report. ngineering and Statistical Research Institute Report of Contract File #24SU.01916-2-L14, Agriculture Canada, Ottawa, Ontario, 214 pp.

asschelein, C.A. (1973). Effect of Beer Bacteria on Beer Flavours. The Brewer's igest, 8, 54.

irt, S.J. (1975). Principles of Microbe and Cell Cultivation. New York. Wiley.

iner, J. (1978). Determination of Total Carbohydrates in Beer. J. Inst. Brew., 84, 222.

THE DEVELOPMENT OF VARIABLE SURFACE REACTORS FOR BIOCONVERSIONS

Paul P. Matteau, National Research Council, Bioenergy Program

and

Christin E. Choma, Queen's University, Kingston, Ontario

ABSTRACT

Two types of variable-surface bioreactors (tapered-bed and radial-flow) have been compared to packed-bed tubular reactors for carrying out bioconversions using immobilized cells. Two length-to-diameter ratios for each reactor type were used and each reactor was evaluated using inert conditions as well as a single-step enzymatic process (cellobiose to glucose) and a multi-step fermentation (glucose to ethanol). Both variable-surface types were found to be as effective as the tubular reactors for the enzymatic hydrolysis of cellobiose to glucose. For the model fermentation of glucose to ethanol with calcium alginate entrapped Saccharomyces cerevisiae cells, use of the tubular reactors resulted in catalyst bed separation and bed compaction with concommitant destruction of the alginate beads and very high pressure drops (160 kPa for a 1 litre bed). On the other hand, none of the variable-surface area bioreactors produced such drastic operational difficulties. The results obtained showed that the variable-surface reactors not only reduced inherent pressure drops due to liquid flow but were also effective in permitting the dissipation of the gas formed during the fermentation reaction.

INTRODUCTION

The advantages associated with the utilization of immobilized enzymes and whole cells for continuous bioconversions have been well documented and have resulted in the initiation of many new immobilization techniques (1). These efforts have, therefore, necessitated the design of bioreactors to be used in conjunction with the biological catalytic particles. By far the most often employed reactor type for use with immobilized cells and enzymes is the packed-bed tubular reactor with the biocatalyst in the form of spheres, chips, disks, sheets, beads or pellets. In such a reactor, there is a steady movement of substrate across the bed of immobilized cells/enzymes in a chosen spatial direction and if the velocity profile is perfectly flat over the cross-section, the reactor is said to operate as a plug-flow reactor (PFR).

Packed-bed reactors have the advantage of simplicity of operation, high mass-transfer rates, and high volumetric reaction rates (for processes which are not inhibited by the substrate concentration). On the other hand, a unique feature of the packed-bed reactor is that there exists a pressure drop or resistance to flow across the reactor. This pressure drop depends on such parameters as bed height, feed rate, feed temperature, composition and catalyst packing density. This feature can result in gas hold-up, channelling of the fluids and compaction of the bed. This drawback is further compounded during the course of certain biological reactions where a substantial amount of gas is produced. The increased pressure drop across the reactor which may result from inefficient gas removal poses an even greater problem as a reduction in the flow of the liquid phase will occur. Increasing the flow merely acts to further increase the pressure differential across the packed-bed and results in additional compaction of the bed and a greater reduction in the hydraulic permeability of the resulting bed. New reactor designs which have been proposed to overcome these problems have included fixed-film reactors with large void volumes (2), suspended-bed reactors (3), inclined tubular reactors with in situ gas/liquid separators (4) membrane reactors, (5), parallel flow reactors (6), and fluidized-bed reactors (7).

In order to maintain the positive features of the packed-bed reactor, two novel reactor geometries have been considered: a radial-flow packed-bed (RFPBR) and a tapered packed-bed reactor (TPBR). In both geometries there is an increase in bed surface area in the direction of flow. This increase in area reduces the fluid velocity through the bed to give a lower pressure drop than that obtained in an equivalent tubular reactor bed. Additionally, there is a reduction in the bed resistance to the flow of gases which are produced during the bioconversion.

The present work has involved the design, constructon and testing of four variable-surface area packed-bed reactors (two radial-flow units and two tapered-hed units). The performance of these reactors was then compared to that of tubular packed-bed reactors for both an single-step enzymatic conversion and for a multiple-step model fermentation using calcium alginate encapsulated whole cells. The enzymatic reaction which was tested was the hydrolysis of cellobiose to glucose using a mycelium-associated activity of Trichoderma harzianum E58. The model fermentation of glucose to ethanol and carbon dioxide was carried out with the yeast Saccharomyces cerevisiae NRC 202076.

METHODS

Reactor Design/Construction

In order to adequately test the performance of the variable-surface reactors relative to tubular reactors two reactors of each type were constructed using two different lenght-to-diameter ratios. Both the tubular and the radial-flow reactors were constructed of acrylic plastic whereas the shorter of the tapered-bed reactors was formed of ABS plastic and the taller unit was constructed of styrene-acrylonitrile. The actual dimension of each of the reactors is presented in Table 1. Each of these reactor geometries represent a range of L/D ratios which were thought to be useful for each reactor type.

All reactors were operated isothermally using either a heat-exchange jacket (tubular reactors) or by immersing the reactors in a constant-temperature water bath. A schematic of the general system under operating conditions is presented in Figure 1. The substrate was fed to the reactor using a peristaltic pump and the effluent was collected for a timed internal in a gas/liquid separator which was found essential for the fermentation. The two solenoid valves (one to a drain and the other to a gas meter/vent) were actuated through a relay controlled by a multiplexer interfaced to a HP85 computer. During the fermentation reaction, the

TABLE 1: Description and Designations of Reactors

Reactor Type/ Designation	Height (cm)	Diameter (cm) Inlet Outlet	
TUBULAR			
TUB 1	49.3	5.1	5.1
TUB 2	87.7	3.8	3.8
TAPERED-BED			
TAP 1	16.0	1.9	15.0
TAP 2	45.0	1.3	7.6
RADIAL-FLOW			
RAD 1	8.2	1.3	12.7
RAD 2	24.7	1.3	7.6

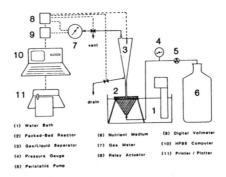

(1) Water Bath

(2) Packed-Bed Reactor   (6) Nutrient Medium   (9) Digital Voltmeter

(3) Gas/Liquid Separator   (7) Gas Meter   (10) HP85 Computer

(4) Pressure Gauge   (8) Relay Actuator   (11) Printer / Plotter

(5) Peristaltic Pump

FIGURE 1: Schematic diagram of system used for
reactor evaluation

evolved gas was monitored by a gas meter electronically coupled to the minicomputer and permitting the on-line evaluation of the efficiency of the conversion.

Immobilization Procedure

The general procedure used to encapsulate the yeast and fungal mycelium involved the separation of the cells from the culture filtrate and resuspension of washed solids in an aqueous 1.5% sodium alginate suspension. This mixture was then homogenzied (food blender) and then extruded under pessure through stainless steel needles into a 4% aqueous solution of calcium chloride. The bead size could be varied by changing the needles which were connected to a 5 litre polycarbonate vessel. The formed beads were normally left for several hours in the calcium chloride solution before use.

Microorganisms

The fungus used for this study was Trichoderma harzianum E58 which had been obtained from the culture collection maintained by Forintek Canada Corporation. Stock cultures were maintained at 4°C and the inoculum culture were grown for 7 days at 30°C on a rotary shaker in Vogel's medium using 2% Solka Floc as the carbon source (8). Normally, four 1-litre flasks containing 400 ml of medium were grown for 5 days before the solids were removed by filtration and then encapsulated in the calcium alginate beads.

The yeast used was Saccharomyces cerevisiae NRC 202076 obtained from the National Research Council. The yeast was grown in static culture without aeration at 30°C in a 5% glucose medium containing the following per litre of solution: 0.5g KH_2PO_4, 0.5g NaCl; 0.35g $MgSO_4$; 20.g $(NH_4)_2SO_4$; and 1.0g of yeast extract. It was determined that under long term serial batch culture reduction of either the salts or the yeast extract concentrations resulted in reduced alcohol production. This medium with the addition of 1.5g of calcium chloride per litre was used for the contiuous fermentations using the immobilized cells. The cells from 2 litre of this medium were normally used for the immobilization step by resuspending them in 8 litres of the sodium alginate suspension and extruding the mixture through needles having a diameter of 1.024 mm. After hardening in the calcium chloride solution, the beads were transferred into 8 litres of aerated 5% glucose medium and maintained at 30°C. The medium was replaced each day and after 4 days of incubation, the beads containing the actively growing yeast cells were wet-packed into the reactors.

RESULTS

Enzymatic Hydrolysis

In order to compare the relative effectiveness of each of the reactors, three reactors were run simultaneously, one tubular, one tapered-bed and one radial-flow. Using calcium alginate beads of 2 mm diameter, it was possible to pack 770 ml (displaced volume) of the encapsulated Trichoderma mycelium in each of the reactors. The results presented are for the taller of each of the tubular and the tapered reactors and the shorter radial reactor.

The dynamic response of each of the reactors was found to be very similar, in that the times required to reach steady-state conversions were comparable and were found to vary directly with the substrate concentration and inversely with the dilution rate which was used. On the average it was found that steady-state conversions could be achieved in approximately 3 hours of operation at 50°C. Typical results for the conversion of cellobiose to glucose using these reactors at a 5mM substrate level are presented in Figure 2. As can be seen the radial-flow reactor gave maximum conversions of approximately 60% whereas the other two reactors achieved 100% conversion at low dilution rates, as would be expected from this type of system. Similar results were achieved with higher substrate concentrations. Once the runs were completed, examination of the radial-flow reactor showed that the calcium alginate bed had shrunk. The decreased effectiveness of this reactor was, therefore, the result of channelling of the substrate through the bed. It was subsequently found that this problem could be overcome by increasing the packing density in the reactor. No such problems were found to occur in the other two reactors. This is likely as a result of there characteristic upflow mode of operation. Additionally, as might be expected, no measureable differences were obtained with either of these reactors at any of the substrate concentrations (5 to 30mM) or dilution rates (0.05 to 3.5 1/h) used.

Experiments with the other reactors indicated that no operational or kinetic advantage could be achieved using the variable-surface reactors for such a bioconversion. The pressure drops obtained across the reactors were consistently low at all dilution rates used and except for the channelling behaviour observed in the radial-flow reactor, the degree of conversion at given dilution rates was consistently equivalent.

Fermentation

The effectiveness of the various reactors for carrying out a gas producing bioconversion was determined in all but the taller of the tubular reactors. Initial flow experiments without reaction had suggested that the use of a tubular reactor with a length-to diameter ratio of 23/1 resulted in unacceptably high pressure drops across the reactor. It was also found that even at low dilution rates the calcium alginate bead bed was sufficiently disrupted by the high linear velocities so as to result in bed separation and subsequent compaction of the bed. Comparative runs with the other reactors were, therefore, carried out by operating two or three reactors under identical loading and operating conditions. After 24 to 48 hours of operation using a 5% glucose medium in order to permit the growth and acclimitization of the immobilized yeast cells, the desired substrate concentration and dilution rate were set and the reactions were allowed to proceed until a steady-state level of conversion had been achieved. Such a process is illustrated in Figure 3 for the short tubular and radial-flow reactors and the tall tapered-bed reactor. The time required to achieve steady-sate conversions varied from 6 to 10 hours at a dilution rate of 0.9 1/h and a substrate concentration of 5%. Generally, the radial-flow reactor achieved maximum conversions more rapidly than either of the upflow reactors. On the other hand, the degree of conversion in the radial-flow reactor was consistently lower.

A major advantage of using the variable-surface reactors which had been predicted and which was borne out during the course of the experiments

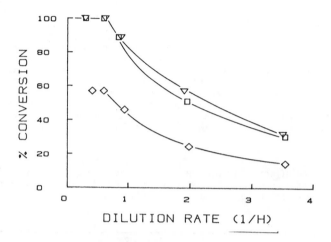

FIGURE 2: Steady-State conversion of 5 mM cellobiose to glucose using the following reactors:

Tall tubular (▽); Tall tapered-bed (▫); Short radial-flow (◇).

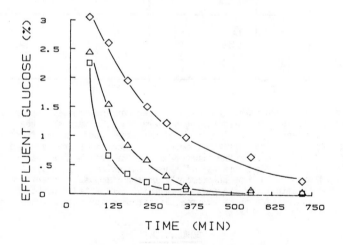

FIGURE 3: Dynamic start-up response of the short tubular reactor (◇), tall tapered-bed reactor (△), and the short radial-flow reactor (▫), operating at a dilution rate of 0.9 h^{-1} and a 5% glucose feed concentration.

was that the pressure drops across these reactors are substantially less than that obtained in equivalent sized tubular reactors. Results presented in Table 2 indicate that the pressure differential in the radial-flow reactor are approximately 5% of those obtained in the tubular reactor and that the tapered-bed reactor gave pressure drops of approximately 50% of those in the tubular reactor. In the uplow reactors, the gas which was produced was entrapped in the alginate beads and caused bed separation to occur. In the tubular reactor this resulted in extreme bed compaction and a serious increase in the pressure drop across the reactor. Once this pressure drop had reached 200 kPa, this reactor was shut down and the bed examined. It was found that the calcium alginate bed had been so severely compressed that a substantial amount of deterioration had occurred.

TABLE 2: The effect of reactor geometry on pressure drop. Pressure Drops and productivities were measures after 10 hours operation on 5% glucose medium at the given dilution rate.

Dilution (h^{-1})	REACTOR					
	Short Tubular		Tall Tapered-Bed		Short Radial-Low	
	Pressure Drop (kPa)	Producti- vity (g EtOH/L /h)	Pressure Drop (kPa)	Producti- vity (g EtOH/L /h)	Pressure Drop (kPa)	Producti- vity (g EtOH/L /h)
0.9	13	18.5	13	18.5	1.3	17.6
1.5	28	16.5	21	22.5	2.7	19.5
2.1	41	23.1	14	33.6	6.0	27.3
2.7	80	25.7	27	35.1	2.7	21.6

Although this problem has received passing mention by most researchers working with such compressible beds, very few solutions have been suggested which will overcome this problem. On the other hand, it was found that under most operating conditions tested, no such large pressure drops were ever achieved in either the tapered-bed or the radial-flow reactors. As can be seen in Table 3 the use of higher concentrations of substrate (8% and 12%) did not lead to unacceptable pressure drops. Longer term experiments with these reactors showed that even after 4 weeks of continuous operation on 8% glucose, little deterioration of the calcium alginate beads was obtained.

CONCLUSIONS

The results obtained from our experiments have shown that, in general, variable-surface reactors are as efficient as tubular reactors for carrying out reactions using packed beds of immobilized cells. A drawback of the radial-flow reactor is that when used with a particle which may shrink, channelling can occur. This can be overcome by either overloading the reactor, or preconditioning the catalyst before loading it into the

TABLE 3: The effect of reactor geometry on pressure drop in the two tapered-bed reactors. Pressure drops and productivities were measured after 16 hours operation at the given dilution rate.

Glucose Feed (%)	Dilution Rate (h^{-1})	REACTORS			
		Short Tapered-Bed		Tall Tapered-Bed	
		Pressure Drop (kPa)	Productivity (g EtOH/L/h)	Pressure Drop (kPa)	Productivity (g EtOH/L/h)
8	0.25	21	9.7	24	10.3
	0.6	17	20.4	39	23.5
	0.75	17	27.7	45	31.8
	1.2	17	35.0	31	40.7
12	0.28	14	15.2	19	16.6
	0.38	10	19.7	12	22.4
	0.54	12	23.7	26	27.0
	0.81	10	28.7	24	34.7

reactor. For gas-producing fermentations using packed beds of particles which pack uniformly, the variable-surface reactors were shown to be effective in reducing intrinsic pressure drops as well as permitting a more facile dissipation of the carbon dioxide which is produced in the reaction. The two designs which are suggested here thus permit the use of packed-bed reactors, with their inherent advantages, for all types of immobilized enzyme reactions whether with compressible or incompressible particles.

ACKNOWLEDGEMENTS

This work was supported by a contract from the Division of Energy, NRC and by a grant from the Department of National Defense.

REFERENCES

(1) Klein, J. and F. Wagner: Appl. Biochem. Bioeng., 4 12 (1983).

(2) Sitton, O.D. and J.L. Gaddy: Biotechnol. Bioeng., 24, 1735, (1980).

(3) Scott, C.D. and C.W. Hancher: Biotechnol. Bioeng., 18, 1393, (1976).

(4) Margaritis, A., P.K. Bajpai and J.B. Wallace: Biotechnol. Lett., 3, No. 11, 613 (1981)

(5) Kan, J.K. and M.L. Shuler: Biotechnol. Bioeng., 29, 217 (1978).

(6) Nojima, S.: Chem. Econ. Eng. Rev., 15, 17 (1983).

(7) Bauer, W.: Abstracts of 34th Can. Cehm. Eng. Conf., 264 (1984).

(8) Matteau, P.P. and J. Saddler: Biotechnol. Lett., 4, 513 (1982).

KINETIC MODELLING AND COMPUTER CONTROL

OF CONTINUOUS ALCOHOLIC FERMENTATION IN THE

GAS-LIFT TOWER FERMENTER

* D. Martin Comberbach, [+]Charles Ghommidh, and John D. Bu'Lock

Weizmann Microbial Chemistry Laboratory, The University of Manchester, M13 9PL, England.

* Present Address: Allelix Inc.
                    6850 Goreway Drive
                    Mississauga, Ontario
                    L4V 1P1, Canada.

[+] Present Address: Laboratoire de Génie Microbiologique
                     Université des Sciences et Techniques
                     du Languédoc, Place E. Bataillon
                     34060 Montpellier Cédex, France.

ABSTRACT

A single stage gas-lift tower fermenter was designed, built and operated under anaerobic conditions to obtain kinetic data needed for the design of a commercial pilot-scale process producing industrial ethanol.

A highly flocculent strain of Saccharomyces uvarum was used to convert glucose to ethanol and carbon dioxide on a continuous basis. The required degree of mixing and the circulation through a gravity-settled quiescent zone was provided by the gas-lift effect, driven by pumped recirculation of a portion of the off-gases. The use of a strongly flocculent yeast strain allowed the reactor to run with virtually total biomass retention and a clear effluent.

Monitoring and control of the reactor was facilitated by the installation of a computer-controlled gas chromatograph which measured the vapour phase ethanol concentration "on-line" through the headspace gas.

Satisfactory operation was maintained over prolonged runs at biomass concentrations up to 150 g.L^{-1} (dry weight), and hydraulic dilution rates greater than 2.5 h^{-1}. The maximum ethanol concentration and productivity attainable were 88 g.L^{-1} and 44.5 g. $^{-1}$.h^{-1}, respectively.

INTRODUCTION

Since the beginning of the century, the ethanol fermentation producing fuel or feedstock was based on the traditional batch process for the manufacture of alcoholic beverages. However, a recent study [1] has shown that significant economic advantages exist in both capital and operating costs for continuous, and continuous with cell recycle, modes compared with the traditional batch route. Since then, various types of laboratory-scale reactors have been used to effect cell recycle during continuous fermentation, the most common being the mechanically-agitated vessel with an attached separator (sedimentation vessel or centrifuge) which returns a concentrated biomass stream to the mixed zone [2,3] and permits a relatively clear effluent to be fed to the distillation columns. The combination of CSTR with gravity separation requires the settler stage to accommodate both floc formation (from yeast dispersed in the CSTR) and actual settling, and such combinations have in practice been limited by the settler efficiency.

The natural ability of yeast cells to flocculate was exploited in the tower fermenter [4,5] developed for the brewing industry. More recently the tower fermenter was employed to produce ethanol for non-beverage uses, not only with a naturally flocculent yeast [6] but also with a previously non-flocculent strain which was induced to flocculate under the physico-chemical conditions with the vessel [7].

The combined effects of CO_2 production and yeast deflocculation in the tower fermenter ensure that operation is midway between a plug-flow (PF) and a fluidized bed reactor (FBR). This imposes design problems in kinetic modelling, location of instrumentation and process control on scale-up. It also means that operating characteristics are sensitive to the distribution of different particle characteristics through the reactor volume and may vary unpredictably. An ideal situation would be to combine the cell retention characteristics of the low-shear tower fermenter with the mixing characteristics of a CSTR in a single stage. The following work indicates how the above criteria were met in a laboratory-scale gas-lift tower fermenter, and how the steady-state kinetic data were obtained to develop an unstructured, mathematical model which was used to predict steady-state behaviour over a wide range of conditions.

MATERIALS AND METHODS

1. Organism and Culture Conditions

Details concerning the organism and culture conditions may be found elsewhere [8].

2. Fermenter Description

A schematic diagram of the gas-lift tower fermenter (GLTF) and its associated apparatus can be seen in Fig. 1.

The fermenter was constructed from Pyrex tubing of 2 mm wall thickness as three separate vessels; the riser 1, the gas cyclone 2 and the sedimentation zone 3. These were linked together using silicone tubing to provide a working volume V_t of 6 L giving a mixed volume V_m (riser etc.) of 4.235 L. The aspect ratio H/D was 14.5:1 and the riser-to-downcomer area ratio A_r/A_d was 7:1. Four 45° ports were

210

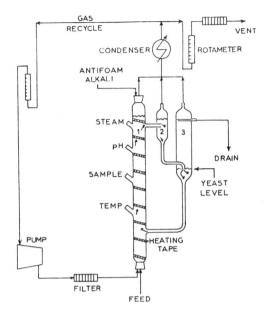

Fig. 1. Apparatus layout of the 6 L GLTF.

located in the riser for instrumentation and sampling. Efflux gas (CO_2)
was removed through a water-cooled condenser, a rotameter and a
pre-sterilized depth filter before being returned to the base of the riser
via a twin-nozzle sparger. A small portion was removed for "on-line"
head-space gas chromatography and the remainder was vented to the
atmosphere. Further details concerning apparatus layout and dimensions
are given elsewhere [9].

3. Biomass Build-Up and Continuous Operation

Biomass build-up from inoculation to continuous run start-up took
approximately 4 days during which time the yeast concentration increased
from 1.5 to 100 g L^{-1}. Accumulation of a working cell density was
facilitated by air sparging (4 L min^{-1}) the pre-sterilized 10% MYGP
solution. After 24 h, continuous feeding was begun using a 5% feed
solution at D = 0.5 h^{-1} (on the mixed volume). When sufficient viable
biomass had been generated, the fermentation was initiated by substituting
the dilute feed with the medium containing the required glucose
concentration and by switching over from air sparge (aerobic) to gas
recycle (anaerobic).

A range of input glucose concentrations was chosen between 15 and 280
g L^{-1}, and dilution rates were selected over the range 0.21-2.83 h^{-1} on
the mixed volume V_m. V_m was estimated geometrically by calculating
the volume of liquid contained within the riser, gas cyclone, downcomer
and a portion of the sedimentation zone (distinguished by a fixed

horizontally drawn line on the external vessel); $V_m/V_t = 0.706$. The arbitrary line corresponded approximately to the level above which the sedimentation zone contained virtually clear liquor and where the occasional buoyant yeast floc could degas before descending back into the mixed zone . An approximate steady-state biomass concentration was maintained by visual estimation of the yeast level in the fermenter and by removal of smaller or larger sample volumes to compensate for any variation. An increase in biomass concentration did not result in a noticeable increase in the height of the sedimenting yeast bed, thus justifying the above V_m/V_t demarcation. The fermentation was terminated when glucose and/or ethanol concentrations remained relatively constant over a duration of several hours.

4. Analytical Methods

Aliquots (100 ml) for the analysis of yeast, glucose and ethanol concentrations were removed from the riser and weir effluent at approximately 2 h intervals. The glucose concentration was determined using an industrial sugar analyser (YSI, OH), and the ethanol concentration was determined by gas chromatography (Perkin-Elmer, Sigma 1B) using AnalaR isopropanol as the internal standard. In some experiments the ethanol concentration was measured by "on-line" headspace gas chromatography by connecting the fermenter head space gas directly to the gas chromatograph and performing the sample injection automatically. Details of this technique are reported elsewhere [10,11].

THEORETICAL ANALYSIS OF THE GAS-LIFT TOWER FERMENTER

A simple mathematical model of the GLTF was proposed based on observations of the hydrodynamic behaviour of the yeast flocs within the reactor. The following reasonable assumptions were made:

(a) The fermenter was divided into two well-defined zones: an adequately mixed zone (which included the riser, the gas cyclone, the downcomer and the lower portion of the sedimentation zone) and an unmixed zone (the upper portion of the sedimentation zone).

(b) Each zone volume remained constant despite changes in substrate feed rate, biomass concentration and floc morphology.

(c) the yeast flocs, the substrate and the product were homogeneously distributed throughout the mixed zone, and the unmixed zone exhibited plug flow characteristics.

(d) Negligible quantities of liquid were lost (as vapour) in the head space gas.

A diagrammatic representation of the fermenter and its equivalent model was constructed (Fig. 2). The shaded region represents the mixed zone.

The total fermenter working volume V_t was thus the sum of the mixed zone volume V_m and the sedimentation zone volume V_s; hence three residence times, the overall hydraulic residence time τ_h, the mixed

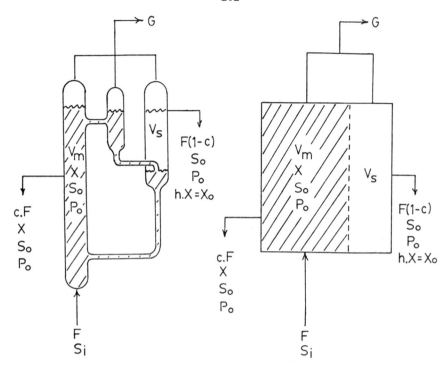

Fig. 2. Diagrammatic representation of the fermenter and its equivalent model.

volume residence time τ_m and the mean biomass residence time τ_x, were defined as follows:

$$\tau_h = V_t/F$$
$$\tau_m = V_m/F$$
$$\tau_x = V_m/cF$$

The mixed zone was by definition the only region in which the substrate, biomass and product concentrations were uniformly distributed; therefore a dilution rate $D = \tau_m^{-1}$, called the dilution rate on the mixed volume, was used in the subsequent analysis.

A material balance on the components of interest gave the following equation for the biomass:

$$(1) \qquad \frac{dX}{dt} = \mu X - cDX - (1-c) DhX$$

The unmixed zone was almost completely devoid of yeast flocs and therefore a negligible quantity of biomass was lost in the weir output (100% biomass retention); thus $hX = 0$, and a biomass dilution rate d could be defined as $d = \tau_x^{-1} = cF/V_m$. Equation (1) therefore became:

(2)
$$\frac{dX}{dt} = \mu X - dX$$

For product and substrate the following equations were written:

(3)
$$\frac{dP_o}{dt} = \nu X - D P_o$$

(4)
$$\frac{dS_o}{dt} = D(S_i - S_o) - qX$$

where μ (h^{-1}) is the biomass specific growth rate, ν (g product (g cells)$^{-1}$h^{-1}) is the specific product formation rate and q(g substrate (g cells)$^{-1}$h^{-1}) is the specific substrate consumption rate. The predicted steady-state behaviour according to the above equations was given by setting the derivatives of biomass, product and substrate concentrations equal to zero:

(5)
$$\frac{dX}{dt} = \frac{dP_o}{dt} = \frac{dS_o}{dt} = 0$$

From equations (2), (3) and (4) respectively, the following equations were derived:

(6)
$$\bar{\mu} = d$$

(7)
$$\bar{P_o} = \frac{\bar{\nu}\bar{X}}{D}$$

(8)
$$\bar{S_o} = S_i - \frac{\bar{q}\bar{X}}{D}$$

where a bar over a symbol represents the steady-state value.

In continuous operation, the biomass dilution rate d was adjusted either by manual sampling or by pumping out from the mixed zone to maintain the steady-state. With total biomass retention (d = 0), it follows from equation (6) that the growth rate μ is equal to zero, a situation possible only under substrate limiting conditions were it not for the presence of a growth-inhibitory product, i.e. ethanol. For this, a certain critical ethanol concentration P_m was defined which completely inhibits biomass growth. It therefore follows that, when d = 0, the steady-state ethanol concentration P_o is equal to P_m.

RESULTS

From the experimental results for a range of dilution rates and substrate input concentrations, three performance regions were identified, corresponding to (a) biomass, (b) product and (c) substrate limitations. Identification of each region was facilitated by plotting the volumetric ethanol productivity DP_o versus the dilution rate D [8]. Biomass limitation was due to the technical limit of fermenter yeast holdup. Product limitation was caused by the kinetic effects of ethanol inhibition on its own production, and substrate limitation arose because insufficient glucose was being fed to the fermenter. With excess substrate, glucose was converted to ethanol at rates dependent on the dilution rate and biomass concentration-viability up to an ethanol concentration of approximately 80 g L^{-1}.

As a consequence of the dependence of biomass concentration and viability on volumetric productivity, an increase in performance for a constant dilution rate and fixed concentration was the result of increases in both biomass concentration and viability up to approximately 150 g L^{-1} and 90% respectively. Since it was never possible to achieve an observed 100% viability using the methylene blue stain method (owing to natural aging), a viability of 90% was presumed to be the maximum possible for this yeast strain. Both μ and ν were calculated by dividing the total quantity of either biomass (samples plus concentration increase) or ethanol produced in 12 h by the mean mass of contained cells within the steady-state period. Plots of μ and ν versus P_o are shown in Fig. 3. Points likely to show substrate limitation ($S_o < 5$ g L^{-1}) were omitted.

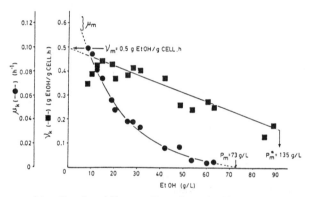

Fig. 3. Specific growth and production rates
vs. ethanol concentration.

An apparent product yield coefficient $Y_{p/s}$ was defined as the quantity of ethanol formed from an observed quantity of glucose. It was found to be completely independent of the ethanol concentration, and its mean value was calculated as 0.425 g ethanol (g glucose)$^{-1}$.

Visual observation of the yeast biomass within the fermenter indicated a doughnut-shaped floc morphology (2-3 mm in diameter, 0.5-0.8 mm thick) which proved to be stable under conditions of varying feed rate and substrate and product concentrations. Both flocculation and sedimentation characteristics remained largely unchanged over several months of continuous operation. Selective yeast retention was maintained even after moderate contamination (e.g. from feed lines), and any infecting organisms were quickly washed out without any subsequent effect on reactor performance.

DISCUSSION

1. The Kinetics of Yeast Growth and Ethanol Production

The most common mathematical approach to kinetic modelling is to use a simple unsegregated [12] model which treats the culture mass as the fundamental variable and ignores the presence of individual cells. This approach reduces the complete growth process to one of simple biological kinetics. Monod [13] was one of the first workers to propose a simple unsegregated model for bacterial growth. It is an unstructured model because it does not account for any physiological variations in the culture mass. The specific growth rate is usually expressed as a function of the concentration S_o of the limiting substrate:

$$(9) \qquad \mu = \mu_m \frac{S_o}{K_s + S_o}$$

With an inhibitor present, the specific growth rate should be extended as $\mu = g(S_o, P_o)$, the characteristic of the function $g(S_o, P_o)$ depending on the nature of the inhibitor. Since ethanol is reported to be a non-competitive inhibitor of both yeast growth and ethanol production, K_s and K_s^* will be unaffected by P_o.

Four equations describing the dependence of μ on P_o have been proposed, all of which are rough approximations of much more complicated effects:
(a) linear [14-16],

$$(10) \qquad \mu_K = \mu_m - k_p^{(1)} P_o$$

(b) exponential [17]

$$(11) \qquad \mu_K = \mu_m \exp\left(-k_p^{(2)} P_o\right)$$

(c) hyperbolic [18-21]

$$(12) \qquad \mu_K = \mu_m \frac{k_p^{(3)}}{k_p^{(3)} + P_o}$$

and (d) parabolic [22],

$$(13) \qquad \mu_K = \frac{\mu_m}{\left(1 + P_o/P_m\right)^{0.5}}$$

Insertion of any of equations (10-13) into equation (9) results in expressions describing both substrate limitation and inhibition by ethanol; e.g. using equation (12).

$$(14) \qquad \mu = \mu_m \left(1 + \frac{k_p^{(3)}}{P_o}\right) \frac{S_o}{k_s + S_o}$$

For $S_o > K_s$, equation (14) reduces to equation (12).

Expressions describing the effects of ethanol inhibition and substrate limitation on the specific growth rate can be readily applied to the specific ethanol production rate. Accordingly, both the specific growth rate and the specific production rate for all the steady-state experiments performed excluding those which were likely to be substrate limited were used to formulate an unsegregated unstructured mathematical model.

Examination of the data describing μ and P_o (Fig. 3) suggested a model of the form:

$$\text{(15)} \qquad \mu_k = \mu_m \left(1 - \frac{P_o}{P_m} \right)^n$$

where P_m, μ_m and n were to be determined. The goodness of fit was insensitive to n between 1.66 and 3.33, so for simplicity a value of 2 was chosen. The numerical values for μ_m and P_m were then calculated from the intercepts on the graph of $\mu^{0.5}$ versus P_o (Fig. 4).

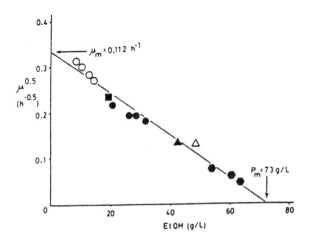

Fig. 4. Verification of specific growth rate model and constant determination.

For ethanol production, an equation of the form:

$$\text{(16)} \qquad \nu_k = \nu_m \left(1 - \frac{P_o}{P_m^*} \right)$$

was chosen, but data describing the variation in ν with P_o were somewhat scattered. However, by assuming stoichiometric catabolism of glucose to equimolar amounts of ethanol and CO_2, a plot of the specific CO_2 production rate ν_G versus P_o should yield a similar relationship to that of ν versus P_o. Using transient data (performed sequentially during the same experimental run), and after corrections accounting for rotameter calibration, headspace gas removal and conversion to standard temperature and pressure, the variation in v_G with P_o was found to be linear Fig. 5).

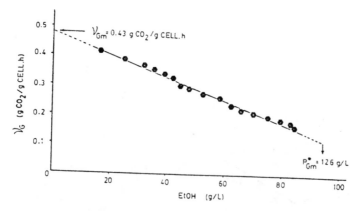

Fig. 5. Verification of ethanol production model via CO_2 efflux.

An approximate value for the substrate saturation constant K_S^* for ethanol production was estimated from a transient experiment during which the substrate from concentration was suddenly changed from 39 to 16 g L^{-1} (Fig. 6).

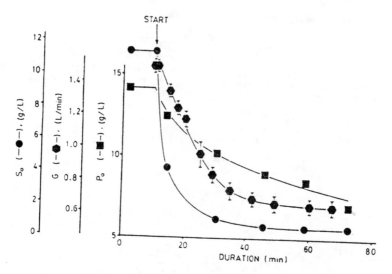

Fig. 6. Transient data at constant D = 2.83 h^{-1}.

The rate of CO_2 evolution was used as a measure of the rate of ethanol production and, by double-reciprocal plotting, a value of K_s^* was obtained from the intercept on the abscissa (Fig. 7).

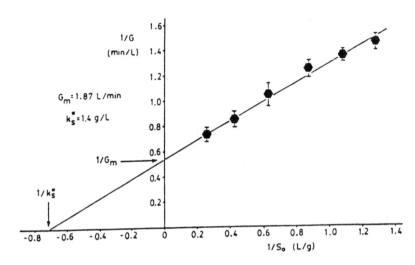

Fig. 7. Verification of specific production rate model and K_s determination.

The equation describing v versus P_o became:

$$(17) \quad V = V_m \left(1 - \frac{P_o}{P_m^*} \right) \frac{S_o}{k_s^* + S_o}$$

A complete mathematical description of yeast growth and ethanol production at steady-state over the experimental operating range of the GLTF could thus be provided by the following equations:

$$(6) \quad \bar{\mu}_k = d$$

$$(7) \quad \bar{P}_o = \bar{v} \bar{X} / D$$

$$(18) \quad \bar{S}_o = S_i - \frac{\bar{P}_o}{Y_{P/S}}$$

$$(15) \quad \mu_k = \mu_m \left(1 - \frac{P_o}{P_m} \right)^n$$

and

$$(17) \quad V = V_m \left(1 - \frac{P_o}{P_m^*} \right) \frac{S_o}{k_s^* + S_o}$$

where the constants were assigned the following numerical values:

$$\mu_m = 0.112 \ h^{-1}$$
$$v_m = 0.5 \ g \ ethanol \ (g \ cells^{-1})h^{-1}$$
$$P_m = 73 \ g \ L^{-1}$$
$$P_m^* = 135 \ g \ L^{-1}$$
$$K_s^* = 1.4 \ g \ L^{-1}$$
$$Y_{P/S} = 0.425 \ g \ ethanol \ (g \ glucose)^{-1}$$
$$n = 2$$

2. Estimation of Unused Substrate

A recent report concerning the industrial production of fermentation ethanol suggests that 50%-60% of the total ethanol production costs are represented by the cost of the substrate [23]. Substrate effluent concentration therefore plays a key role in process economics and must be predictable over the whole operating range. The above mathematical model was used to simulate the fractional unused substrate concentration S_o/S_i in the effluent from a fermenter containing 100 g L^{-1} of 90% viable yeast. Curves with $P_o < P_m$ were sigmoidal as D approached zero (Figs 8,9),

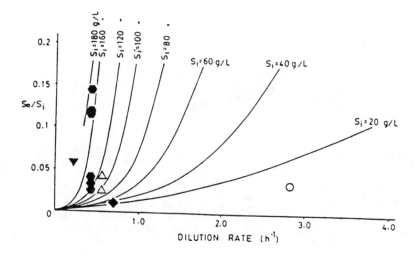

Fig. 8. Simulation of fractional unused substrate vs. dilution rate (biomass concentration, 100 g L^{-1} (90% viable); $v_m = 0.5$ g ethanol (g cells)$^{-1}$h^{-1}; $Y_{P/S} = 0.425$; $P_m = 73$ g L^{-1}; $P_m^* = 135$ g L^{-1}; $K_s^* = 1.4$ g L^{-1}; experimental data are shown for comparison.

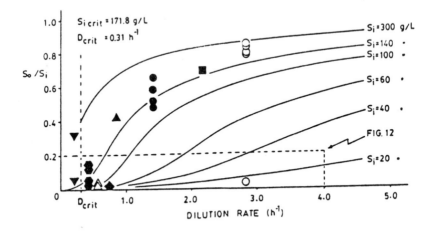

Fig. 9.   Simulation of fractional unused substrate vs. dilution rate
(biomass concentration, 100 g L^{-1} (90% viable); v_m = 0.5 g
ethanol (g cells)$^{-1}$ h^{-1}; $Y_{p/s}$ = 0.425; P_m = 73 g L^{-1};
P_m^* = 135 g L^{-1}; K_s^* = 1.4 g L^{-1}; experimental data are
shown for comparison.

and within this limit the fermentation was stable. However, on reaching
P_m (supplied by an input substrate concentration equal to S_{icrit}),
growth ceased, and a region of instability developed for all ethanol
concentrations between P_m and P_m^*. Single-stage steady-state
continuous fermentation was impossible at ethanol concentrations exceeding
P_m. As T_m was reduced, steady-state fermentation became possible only
at the expense of higher effluent substrate concentrations (biomass
limitation). Experimental increases in biomass concentration and/or
percent viability reduced the fraction of effluent substrate.

3. Fermenter Performance: Experiment Compared with Theory

The mathematical model was used to compare the theoretical and practical
performances of the GLTF by calculating values for v, D and DP_o (eqns. (7)
and (17)) for various steady-state ethanol concentrations between zero and
P_m. The calculation was based on a biomass concentration of 100 g L^{-1}
(90% viable) with the assumption that the substrate was non-limiting. Two
theoretical curves were obtained (Figs. 10,11), the shape of which
corresponded to the experimental performance figures obtained previously
[8].

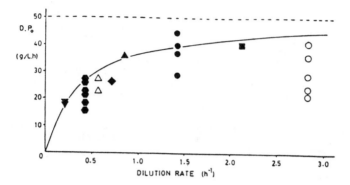

Fig. 10.   Theorectical curve of ethanol productivity vs. dilution
rate: experimental points are shown for comparison.

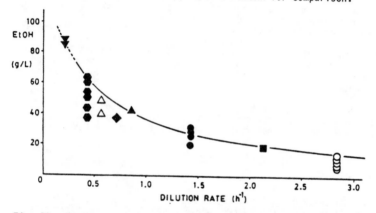

Fig. 11.   Theoretical curve of ethanol concentration vs. dilution
rate: experimental points are shown for comparison.

The steady-state experimental points from all the fermentations are shown
for comparison. The broken line in Fig. 11 indicates the maximum
theoretical productivity attainable in this system, a figure impossible to
achieve in practice since $D \to \infty$ when $P_o \to 0$. The broken line in Fig. 12
indicates the theoretical onset of instability caused by the generation of
a concentration $P_o > 73$ g L^{-1}.

4. Computer Control

For continuous, steady-state operation of the GLTF, the yeast
concentration must be maintained at a constant level either by manual
sampling or by pumping out biomass at the same rate as it is formed,
i.e.:

(6) $\bar{\mu} = d$

Since a direct, on-line sensor for biomass concentration did not exist, the growth rate could only be calculated indirectly, and in this particular instance, the most accessible on-line measurement was the ethanol concentration. This was measured as previously described [10,11], and related to the specific growth rate through equation 15. In practice, the feedback control loop could maintain a steady-state by activating a relay-switched peristaltic pump. This removed biomass from the mixed zone at a rate sufficient to maintain the setpoint effluent ethanol concentration at any value from zero to 70 g/L.

Successive refinements to the control program eventually resulted in a feedback control loop which could maintain an effluent ethanol concentration to within ±3 g/L [24].

CONCLUSIONS

Based on observations of the yeast settling behaviour within the GLTF, a mathematical model was proposed to predict the necessary operating conditions for continuous steady-state ethanol production. Yeast growth and ethanol production kinetics were modelled by two empirical double-constant equations which were valid over a range of ethanol concentrations from 0 to 73 g L^{-1}. The mathematical models of the GLTF and the process kinetics were combined to predict the substrate and product effluent concentrations over a wide range of glucose input concentrations and dilution rates. The predicted parameters were in agreement with the experimental data obtained.

Monitoring and control of the fermenter was facilitated by the installation of a computer-controlled gas chromatograph which measured the vapour phase ethanol concentration 'on-line' through the headspace gas.

REFERENCES

1. G.R. Cysewski and C.R. Wilke, Biotechnol. Bioeng. 20 (1978) 1141.
2. G.R. Cysewski and C.R. Wilke, Biotechnol. Bioeng. 19 (1977) 1125.
3. E.J. Del Rosario, K.J. Lee and P.L. Rogers, Biotechnol. Bioeng. 21 (1979) 1477.
4. M.G. Royston, Process Biochem. July (1966) 215.
5. R.N. Greenshields and E.L. Smith, The Chemical Engineer 249 (1971) 182.
6. J.G. Prince and J.P. Barford, Biotechnol. Letts. 4 (4) (1982) 263-268.
7. J.G. Prince and J.P. Barford, Biotechnol. Letts. 4 (10) (1982) 621-626.
8. D.M. Comberbach and J.D. Bu'Lock, Biotechnol Lett. 6 (2) (1984) 129.
9. D.M. Comberbach, Continuous alcoholic fermentation: A study of ethanol production using a novel gas-lift reactor, Ph.D. Thesis, University of Manchester, October 1981.
10. D.M. Comberbach and J.D. Bu'Lock, Chromatogr. Newsl., 10 (1) (1982) 19.
11. D.M. Comberbach and J.D. Bu'Lock, Biotechnol. Bioeng. 25 (1983) 2503.
12. H.M. Tsuchiya, A.G. Fredrickson and R. Aris, Adv. Chem. Eng., 6 (1966) 125.

13. J. Monod, Ann. Inst. Pasteur, 79 (1949) 390.
14. C.N. Hinshelwood, The Chemical Kinetics of the Bacterial Cell, Clarendon, Oxford, 1952.
15. D.O. Zines and P.L. Rogers, Biotechnol. Bioeng., 12 (1970) 561.
16. T.K. Ghose and R.D. Tyagi, Biotechnol. Bioeng., 21 (1979) 1401.
17. S. Aiba, M. Shoda and M. Nagatani, Biotechnol. Bioeng., 10 (1968) 845.
18. S. Aiba and M. Shoda, J. Ferment. Technol., 17 (12) (1969) 790.
19. N.B. Egamberdiev and N.D. Ierusalimskij, in I. Malek and Z. Fencl (eds.), Continuous Culture of Microorganisms, Academia Praha, Prague, 1969, p. 517.
20. M. Novak, P. Strehaiano, M. Moreno and G. Goma, Biotechnol. Bioeng., 23 (1981) 201.
21. M. Nagatani, M. Shoda and S. Aiba, J. Ferment. Technol., 46 (1968) 241.
22. C.D. Bazua and C.R. Wilke, Biotechnol. Bioeng. Symp., 7 (1977) 105.
23. J.K. Paul (ed.), Ethanol from biomass; comparative economic assessment, Ethyl Alcohol Production and Use as a Motor Fuel, Noyes, Park Ridge, NJ, 1979.
24. J.D. Bu'Lock, D.M. Comberbach, C. Ghommidh and P.D. Williams, Chem. and Ind., 12 (1984) 432.

ACKNOWLEDGMENTS

The authors are indebted both to the Science and Engineering Research Council, London, for the provision of a research grant, and to Sim-Chem Ltd., Cheadle Hulme, Cheshire, England, for their interest and financial support.

NOMENCLATURE

c	fraction of the substrate volumetric feed removed from the mixed zone
d	biomass dilution rate (h^{-1})
D	dilution rate on mixed volume (h^{-1})
D_{crit}	critical dilution rate on mixed volume (h^{-1})
F	substrate volumetric feed rate (1 h^{-1})
G	gas efflux (1 min^{-1})
G_m	maximum gas efflux (1 min^{-1})
h	fraction of the mixed zone biomass in the weir outlet.
$K_p^{(1)} - K_p^{(3)}$	biomass growth inhibition constants (g l^{-1})
K_s, K_s *	substrate saturation constants for yeast growth and ethanol production respectively (g l^{-1})
P_{Gm}*	ethanol concentration at which $v_G = 0$ (g l^{-1})

P_m	ethanol concentration at which $\mu = 0$ (g l^{-1})
$P_m{}^*$	ethanol concentration at which $\nu = 0$ (g l^{-1})
P_o	outlet ethanol concentration (g l^{-1})
q	specific substrate consumption rate (g substrate (g cells)$^{-1}$h^{-1})
S_i	inlet glucose concentrations (g l^{-1})
S_{icrit}	critical inlet glucose concentration (g l^{-1})
S_o	outlet glucose concentration (g l^{-1})
t	time (h, min or s)
V_m	mixed fermenter volume (l)
V_s	sedimentation zone volume (l)
V_t	total fermenter working volume (l)
X	biomass concentration in the mixed zone (g l^{-1})
X_o	outlet biomass concentration from the weir (g l^{-1})
$Y_{p/s}$	apparent yield coefficient of the production substrate (g ethanol (g glucose)$^{-1}$)
μ	yeast specific growth rate (h^{-1})
μ_k	yeast specific growth rate neglecting substrate limitation (h^{-1})
ν	ethanol specific production rate (g ethanol (g cells)$^{-1}$h^{-1})
ν_G	CO_2 specific production rate (g CO_2 (g cells)$^{-1}$h^{-1})
ν_{Gm}	maximum CO_2 specific production rate (g ethanol (g cells)$^{-1}$h^{-1})
ν_k	ethanol specific production rate neglecting substrate limitation (g ethanol (g cells)$^{-1}$h^{-1})
ν_m	maximum ethanol specific production rate (g ethanol (g cells)$^{-1}$h^{-1})
τ_h	overall hydraulic residence time (h)
τ_m	mixed volume residence time (h)
τ_x	mean biomass residence time (h)

ETHANOL PRODUCTION BY ADSORBED CELL BIOREACTORS

Andrew J. Daugulis and Donald E. Swaine
Department of Chemical Engineering
Queen's University
Kingston, Ontario   K7L 3N6

Abstract

The immobilization of whole cells by adsorption onto surfaces, and the use of these cells in packed bed bioreactors is an effective strategy for the continuous production of ethanol. We have immobilized whole cells of Saccharomyces cerevisiae and Zymomonas mobilis onto ion exchange resin beads, and have successfully operated bioreactors ranging in size from 160 mL to 15.7 L capacity. The results of this work and the general considerations of ethanol productivity, ethanol concentration, scale of operation, and long term stability for immobilized cell systems are discussed. A variety of factors pertaining to the use of immobilized cell bioreactors for the production of microbial products other than ethanol are presented.

Introduction

The attachment of microbial cells to surfaces is a widely known phenomenon and, depending on the application, can be viewed as a positive or negative occurrence. It is seen as an undesirable situation when process equipment (e.g. a heat exchanger) becomes "fouled", or when the head space and baffles of a stirred tank fermentor accumulate "wall growth". Surface growth of microbial cells has also been effectively used, however, and has seen extensive application in the treatment of wastewater (e.g. trickling filters and rotating biological contactors).

Recent interest in the immobilization of whole microbial cells and the use of these "fixed" cells in continuous-flow bioreactors has prompted us to examine the adsorption of cells onto surfaces as an alternative strategy to the plethora of studies that have been conducted using the so-called gel-entrapment procedure. As Chibata's recent and excellent review article has pointed out [1], the entrapment of cells suffers from two serious difficulties: mass transfer limitations, and the inability to re-use the carrier. Immobilization of whole cells by adsorption or carrier-binding does not have these drawbacks and, as long as cells can be prevented from leaking away from the carriers, this technique possesses a number of distinct advantages [1].

A review article by Durand and Navarro [2] has summarized earlier work on the adsorption of pure cultures of cells onto a variety of carriers (e.g. wood, brick, P.V.C., ion exchange resins), and the use of these cells for the production of a number of metabolites. Other workers [3,4,5] have examined the adsorption of yeast cells onto wood, unidentified carriers and hollow fibers for the production of ethanol. We have investigated the ability of a number of carriers (ion exchange resins, ceramic chips, activated charcoal) to adsorb cells of Saccharomyces cerevisiae and Zymomonas mobilis, and have developed extremely high productivity ethanol bioreactors using these immobilized microorganisms [6,7,8]. We are also examining the immobilization (by adsorption) of wild-type and genetically-modified cells for the production of high-value polypeptides.

This paper summarizes our cell adsorption work in relation to a number of fundamental as well as practical considerations dealing with immobilized cell bioreactors: cell loadings (initial and ultimate), bioreactor performance (productivity and stability), and application to the production of a variety of microbial products.

Cell Loadings

Perhaps the most obvious advantage in immobilizing whole cells is the resulting ability of a continuous-flow fermentation system to operate at high liquid throughputs while maintaining a high cell concentration even at dilution rates exceeding "washout". This situation might otherwise only be achieved by incorporating a recycle or integrated-settler arrangement. The literature adequately reports immobilized cell systems (entrapped and adsorbed) in which high cell concentrations are maintained at dilution rates which greatly exceed the maximum specific growth rate of the microorganisms in question.

A few early studies on the adsorption of microbial cells to surfaces and the use of these cells for the production of ethanol have reported cell loadings in terms of initial specific adsorption levels (expressed as mg DW cells/g of carrier). For yeast cells, initial specific adsorption levels of 248 mg/g for wood chips [2], 188 mg/g for wood chips [3], and 380 mg/g for some unidentified carrier [4] have been reported. Although useful as a qualitative comparison involving different carriers, data reported in this fashion do not take into account the bulk density of the carrier material. It might therefore be more appropriate to report initial adsorption data in terms of the effective reactor cell concentration (g DW cells/L reactor volume) as this is, in practical terms, the more relevant figure.

As an example of the different picture that emerges when initial cell adsorption data are reported in this manner, Tables 1 and 2 show data for the attachment of cells of S. cerevisiae and Z. mobilis on a variety of carriers. For Z. mobilis, ion exchange resin IRA-938 results in a specific adsorption value which is more than five times greater than for activated charcoal. The effective cell concentration in this case, however, is only about twice as high for the resin. Thus, although the resin is still superior from a cell attachment standpoint regardless of how attachment is expressed, its advantage is less pronounced, and other factors (such as cost) may influence the decision as to which material is "best".

Table 1 : Initial cell loadings for Z. mobilis on various carriers

PACKING MATERIAL	SPECIFIC ADSORPTION (mg DW cells/g packing)	TOTAL ADSORPTION (g DW cells/L reactor volume)
Activated carbon	7.0	2.6
Ceramic chips	1.5	1.6
Ion exchange resins (Gel type)		
IRA-68 (WB , RH)	< 1	-
MIXM-614 (SA/SB , I)	< 1	-
IRA-400 (SB , RH)	1.4	0.6
DOWEX 1-X8 (SB , D)	5.5	2.0
Ion exchange resins (Macroreticular type)		
IRA-958 (SB , RH)	1.3	0.3
IRC-505 (WA , RH)	3.7	1.4
XE-313 (SB , RH)	11.1	1.9
XE-352 (SB , RH)	31.5	5.3
IRA-938 (SB , RH)	36.0	5.7

SB Strongly basic (anionic) WB Weakly basic (anionic)
SA Strongly acidic (cationic) WB Weakly acidic (cationic)

RH Rohm and Haas resin
I Ionac Chemical Co. resin
D Dow Chemical Co.

Table 2 : Initial cell loadings for S. cerevisiae on various carriers

PACKING MATERIAL	SPECIFIC ADSORPTION (mg DW cells/g packing)	TOTAL ADSORPTION (g DW cells/L reactor volume)
Activated carbon	< 1	-
Ceramic chips	< 1	-
Ion exchange resins (Gel type)		
IRA-400 (SB , RH)	< 1	-
IRA-458 (SB , RH)	3.0	.74
IRA-68 (WB , RH)	10.9	3.08
Ion exchange resins (Macroreticular type)		
IRA-900 (SB , RH)	< 1	-
IRA-904 (SB , RH)	4.0	1.02
IRA-958 (SB , RH)	5.3	1.11
IRA-938 (SB , RH)	46.8	6.64
XE-352 (SB , RH)	128	16.1

SB Strongly basic (anionic) WB Weakly basic (anionic)

RH Rohm and Haas resin

More important than initial cell loadings, of course, is the immobilized cell concentration in the bioreactor during operation. For adsorbed cell systems in the simplest case, once all potential adsorption sites are covered, there will be no net change in cell concentration from the initial level (unless reduced by hydrodynamic forces). We have found this to be the case with adsorbed cells of S. cerevisiae (Figure 1) -- an increase in reactor throughput after the establishment of a steady state results in a decreased product concentration. (In order to confirm that the immobilized cell concentration in the reactor in this instance did not substantially change with time, calculations were performed based on specific ethanol production rates for S. cerevisiae -- these calculations revealed that cell concentrations during operation were approximately the same as the concentration immediately after initial adsorption).

Our work with Z. mobilis [7,8] in adsorbed cell bioreactors has proven to be somewhat more complex. Although initial adsorption studies (Table 1) had again identified certain ion exchange resins to be effective in immobilizing whole cells, it was found that when adsorbed cell bioreactors using Z. mobilis were operated continuously, the reactor cell concentration increased substantially with time. Figure 2 shows that several increases in reactor throughput after an initial steady state had been reached resulted in eventual returns to high effluent product concentrations. This result along with visual and microscopic examinations, as well as electron microscopy studies showed several important aspects of adsorbed cell Z. mobilis systems. First, under the conditions that were used, large numbers of filamentous forms of Z. mobilis were produced. Second, the surfaces of the ion-exchange resin beads underwent extensive microbial overgrowth as filamentous cells covered surfaces and pores completely. Third, the overgrowth was sufficiently great to bridge the gaps between adjacent beads and substantially reduce void spaces. Estimates of immobilized cell concentrations in the reactors were made using two methods -- assumed specific ethanol production rates for Z. mobilis, and mass balances in conjunction with experimentally-determined yield coefficients. These calculations substantiated the observations of increasing immobilized cell concentrations during bioreactor operation. Depending on the calculation method employed, immobilized cell concentrations in the range 50-200 g DW/L (based on total reactor volume) were ultimately reached in these systems.

The electron micrographs shown in Figure 3 reveal a number of interesting features concerning the attachment of yeast and bacterial cells to microporous ion exchange resin beads. Figure 3(a) shows a whole resin bead. Beads studied ranged in diameter from 400 to 500 μm and their large surface area is due to their very porous structure. Figures 3(b) & (c) show that attachment of cells occurs both at sites on the outer surface of the resin and in the pores. Bridging of two beads by filaments of Z. mobilis is seen in Figure 3(d), an example of problems encountered in operating adsorbed cell bioreactors.

In summary, the potential of adsorbed cell bioreactors depends to a large extent on the concentration of cells "fixed" within the system. In evaluating potential carriers, care must be taken in interpreting adsorption data, and expressing the data in a meaningful fashion. In addition it is important to recognize that immobilized

229

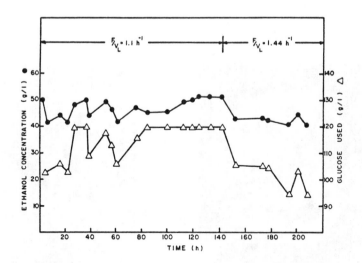

FIGURE 1. Continuous operation of 1.47 L bioreactor with immobilized
S. cerevisiae and influent glucose concentration of 120 g/L. [6]
Dilution rate is expressed on the basis of reactor liquid volume,
V_L = 0.39 L (27%).

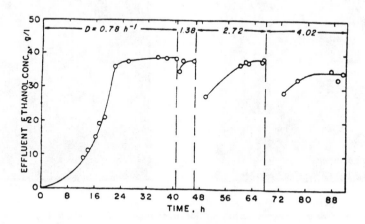

FIGURE 2. Continuous operation of 161 mL bioreactor with immobilized Z. mobilis
and influent glucose concentration of 100 g/L. [7]
Dilution rate (D) is expressed on the basis of total reactor volume.

230

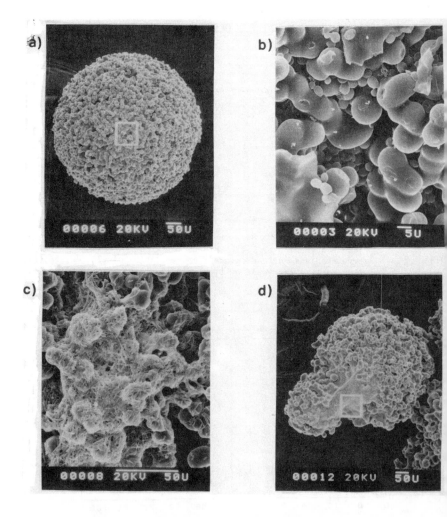

FIGURE 3. Electron micrographs, a) Ion exchange resin bead; b)&c) Closeup of resin bead showing surface and pore attachment of $\underline{S}$. cerevisiae and $\underline{Z}$. mobilis respectively; d) Two resin beads bridged by filaments of $\underline{Z}$. mobilis [6,7].
Bar represents a length of either 50 or 5 microns.

cell concentrations during operation of an adsorbed cell system may differ greatly from the value obtained after initial loading. Either a decrease in bioreactor cell concentration may occur due to hydrodynamic forces which strip cells from surfaces [3], or an increase in cell concentration may occur as a result of morphological and/or surface phenomena [7,8]. In the latter case, Chibata's primary reservation about adsorbed cell systems (cell leakage away from the carrier) appears not to be justified.

Bioreactor Performance

The criterion that many workers have used in comparing the performance of immobilized cell bioreactors for ethanol production has been ethanol productivity. Although from a fundamental standpoint it is tempting to report ethanol productivity on the basis of reactor void volume (thus relating dilution rate to liquid residence time or space velocity), engineering (capital cost) considerations require that productivity be based on total reactor volume. On this basis, the highest ethanol productivities reported in the literature for immobilized cell bioreactors are slightly greater than 100 g/L-h, which compares very favourably to conventional stirred tank systems (≈ 10 g/L-h). Our adsorbed cell Z. mobilis systems have been able to consistently achieve ethanol productivities in excess of 100 g/L-h (with a maximum of 141 g/L-h).

Several other very important factors (besides productivity), however, need to be taken into account in assessing or comparing bioreactor performance. These include final product concentration, demonstrated performance at a reasonable scale, and long term operation/stability. The first of these factors (product concentration) is important in ethanol production, as product recovery costs are significant. Since ethanol productivity is substantially reduced as the product concentration increases (due to a depressed reaction rate), ethanol productivities should be compared on the basis of "realistic" (7-10 % w/v) ethanol concentrations. (Our work has been restricted to final ethanol concentrations in the range 4-5 %.)

The second factor (scale of operation) has to a large extent not been addressed by many workers. Assessment of immobilized cell bioreactor performance has in many cases been made on the basis of laboratory-scale systems ranging in size from less than 100 mL to a few litres in capacity. In evaluating performance for potential industrial applications it will in many cases be necessary to address the attendant scale-up problems relating to heat transfer, gas and liquid hydrodynamics, pressure drop, and any other factors peculiar to each system (e.g. bead compression and fluidization in the case of bead-entrapped systems). We have successfully operated adsorbed cell Z. mobilis bioreactors [8] at the small pilot scale (to 15.6 L capacity), and have successfully dealt with a variety of the aforementioned scale-up considerations.

The final factor (long term operation) has been successfully addressed by a number of workers for a variety of immobilized cell ethanol systems. Our work with adsorbed yeast cells has indicated that long term operation is feasible, although the overgrowth difficulties associated with adsorbed Z. mobilis cells, at least initially, precluded bioreactor operation in excess of 3 or 4 days.

These difficulties have largely been overcome, and continuous operation for several weeks now appears to be feasible.

A recent paper [9] has presented a fairly balanced assessment in reporting on ethanol production in an immobilized cell (alginate bead) system. The system consisted of a 1000 L bioreactor and was capable of producing an 8 % (w/v) ethanol stream for over 4000 L at an ethanol productivity of 20 g/L-h. It is important that workers in the field address all four factors (ethanol productivity, final product concentration, scale of operation, and long term stability) and not rely on a single criterion (e.g. productivity or ethanol concentration) to promote a particular system. Since trade-offs will be inevitable (i.e. Is it "better" to produce 8 % ethanol at 20 g/L-h, or 4 % ethanol at 40 g/L-h?), a quantitative ranking of these factors will be necessary, in which capital costs (largely productivity-related) and operating costs (largely product-concentration-related) are rigorously factored for a particular plant scale. Subjective rankings and continued reliance on single criterion performance assessments are of limited value.

Microbial Products Other Than Ethanol

Immobilized cell systems have, for the most part, been used for the production of ethanol, usually either by yeast cells or by the bacterium Z. mobilis. The ethanol fermentation is arguably a fairly straightforward fermentation involving a single growth-associated product, fairly simple morphological forms (i.e. unicellular), and non-aerated conditions. Other fermentations important in the "bioenergy" area are those used in the production of the so-called "power solvents" (acetone, butanol, butanediol). Still other fermentation processes may be envisaged which use the products of cellulose hydrolysis and yield high-value metabolites. Fermentation technology involving immobilized cells will need to undergo substantial developments and will have to surpass the rudimentary beginnings that have been made with immobilized cell ethanol fermentations. Factors that will need to be addressed include:

1. Immobilized cell aerobic fermentations. Many high-value fermentation products are produced under aerated conditions. Factors that will need to be considered include mass (oxygen) transfer (and the inevitable choice between adsorbed and entrapped cell systems), as well as bioreactor hydrodynamics and pressure drop problems under gassed conditions.

2. Complex kinetics. Non-growth-associated products (secondary metabolites) will require special reactor configurations (e.g. staged, plug-flow, etc.) in order to produce these materials in continuous-flow systems.

3. Hydrodynamics and modelling in packed bed bioreactors. Many immobilized cell bioreactors operate as packed beds, and some understanding of the hydrodynamics in such systems (particularly if aerated, or when gas is evolved) is a precondition for adequate scale-up. This knowledge will also facilitate optimization procedures such as multiple feeding strategies.

The utilization of adsorbed cells for ethanol production is an effective and competitive fermentation strategy. The large-scale potential of this technique (as well as other immobilization techniques) must be scrutinized carefully in a rigorous and balanced way. It is also important to recognize that many microbial products have a higher intrinsic value than ethanol, and the development of immobilized cell systems should perhaps be viewed in this larger context.

References

1. Chibata, I., Tosa, T. and Fujimura, M., "Immobilized Living Cells", Annual Reports on Fermentation Processes, 6, 1 (1983).

2. Durand, G. and Navarro, J.M., "Immobilized Microbial Cells", Process Biochem., Sept. 1978, p. 14

3. Moo-Young, M., Lamptey, J. and Robinson, C.W., "Immobilization of Yeast on Various Supports for Ethanol Production", Biotechnol. Letters, 2, 541 (1980)

4. Ghose, T.K. and Bandyopadhyay, K.K., "Rapid Ethanol Fermentation in an Immobilized Yeast Cell Reactor", Biotechnol. Bioeng., 22, 1489 (1980)

5. Lee, J.H., Pagan, R.J. and Rogers, P.L., "Continuous Simultaneous Saccharification and Fermentation of Starch Using Zymomonas mobilis", Biotechnol. Bioeng, 25, 659 (1983)

6. Daugulis, A.J., Brown, N.M., Cluett, W.R. and Dunlop, D.B., "Production of Ethanol by Adsorbed Yeast Cells", Biotechnol. Letters, 3, 651 (1981)

7. Krug, T.A. and Daugulis, A.J., "Ethanol Production Using Zymomonas mobilis Immobilized on an Ion Exchange Resin", Biotechnol. Letters, 5, 159 (1983)

8. Daugulis, A.J., Krug, T.A. and Choma, C.E.T., "Filament Formation and Ethanol Production by Zymomonas mobilis in Adsorbed Cell Bioreactors", Biotechnol. Bioeng., (in press)

9. Nagashima, M., Azuma, M. Noguchi, S., Inuzuka, K. and Samejima, H., "Continuous Ethanol Fermentation Using Immobilized Yeast Cells", Biotechnol. Bioeng., 26, 992 (1984)

Early Developments of a Novel Fermenter-Purifier

by

D.L. Mulholland
Ontario Research Foundation

ABSTRACT

Originally conceived as a low-energy alternative to distillation as a means of separating ethanol from fermentation beers, the subject of this presentation will be described from its earliest embodiment as an ethanol purifier to its latest role as an ethanol fermenter. The principle on which the system operates is that of gas stripping of the volatile components of the liquid contents of the apparatus. In our case these were ethanol and water. The vessel employed was a cylindrical column fitted with a draft tube. Studies of the kinetics of the system were carried out with a model substrate containing ethanol and water only. Results of these studies will be presented.

Later experiments in which ethanol is produced in situ are described. The ethanol is therefore produced in and removed from the same vessel by volatilisation at temperatures around 40°C. The microorganism employed was Saccharomyces cerevisiae ATCC 4126.

INTRODUCTION

The development of fuel ethanol technology to replace petroleum-based technology is being hindered by unfavourable economic and energy balances. Currently, the attention of many scientists and engineers is being focused on the two areas of perceived sensitivity, viz., alcohol separation and waste treatment. Additionally, in the Canadian context, the utilisation of lignocellulosic feedstocks is also important but the work discussed here is concerned with the first problem area. Alternatives to energy-intensive distillation of dilute beers as a means of separation include vacuum distillation[1], membrane processes[2], adsorption processes[3,4] (e.g. molecular sieves), absorption[5] (solvent extraction), supercritical CO_2 extraction[6] and others. Intensive work is being undertaken in all of these and other areas.

A further alternative is the possibility of low temperature, and, therefore, low energy, gas stripping of the beer. Here, an inert gas is sparged into the liquid, becomes saturated with alcohol vapour and is later chilled and the condensate collected. In order to increase the gas-liquid contact[7] an apparatus has been designed with an internal baffle system to set up a cyclic liquid flow pattern (Figure 1). The baffle is an open cylindrical tube, up the centre of which the gas passes. Voidage in the liquid due to expansion of the gas and the upward movement of the gas bubbles coupled by frictional forces to the liquid cause a circulation pattern up the central tube and down the annular space. Gas is also recycled in the liquid. This increases gas residence time at high gas flowrates.

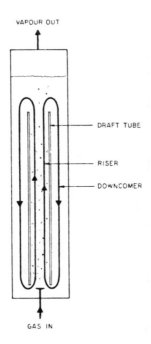

FIGURE 1.   TYPICAL CONFIGURATION OF GAS LIQUID
CONTACTING DEVICE WITH DRAFT TUBE

The vapour phase is enriched in alcohol by virtue of the greater fugacity of the ethanol over water. The relation between the alcohol concentration in the vapour phase and that in the liquid phase is illustrated by the vapour-liquid equilibrium curve. Most data in the literature has been obtained at the boiling point and is applicable to distillation systems (see Figure 2)[8]. To our knowledge, none was available in the 40-60°C range in which we were to operate. The acquisition of these data was one objective in the work described here. Further work to obtain kinetic data and the influence of temperature was carried out.

236

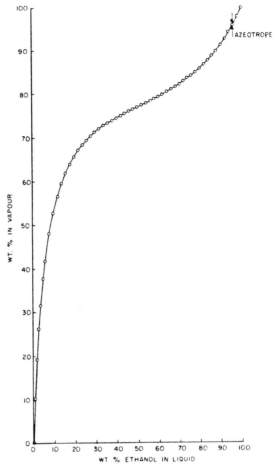

FIGURE 2.   VAPOUR-LIQUID EQUILIBRIUM CURVE FROM LITERATURE

EXPERIMENTAL

A gas liquid contacting device based on the principles described above was constructed according to Figure 3. Water cooled laboratory glass condensers were fitted and residual volatiles in the gas stream were analysed by a long path length gas phase infrared spectrometer. Feedstock and product analyses were carried out by gas chromatography.

The apparatus was charged and tested and found to operate satisfactorily. Air was used as the stripping gas. Except where noted otherwise, data was obtained at 40°C.

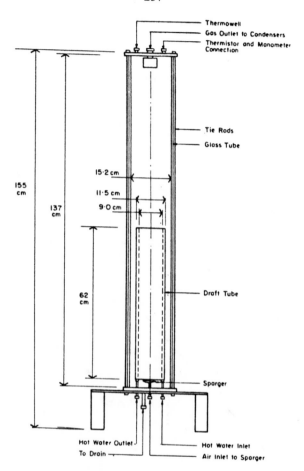

FIGURE 3.   EXPERIMENTAL APPARATUS FOR LOW COST
RECOVERY OF ETHANOL.

RESULTS AND DISCUSSION

Concerns regarding the lack of attainment of equilibrium at high flowrates led us to check the influence of gas flow on the concentration and rate of collection of the product. Results are shown in Figures 4 and 5. It was found that within the limits of experimental accuracy, the concentration of alcohol in the vapour phase and in the condensate was constant for gas flowrates in the range 0.1 to 2.0 VVM (volumes per minute). The rate of volatilisations and of condensate collection increased linearly with gas flowrate.

238

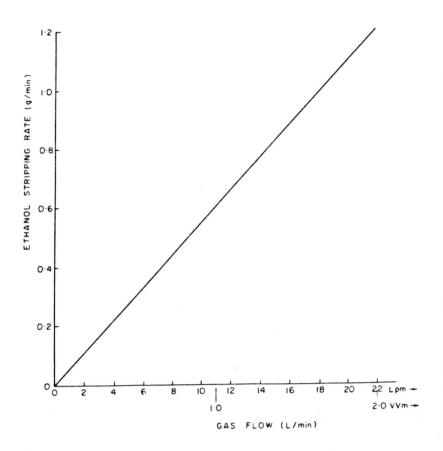

FIGURE 4.   ETHANOL STRIPPING RATE vs GAS FLOW

239

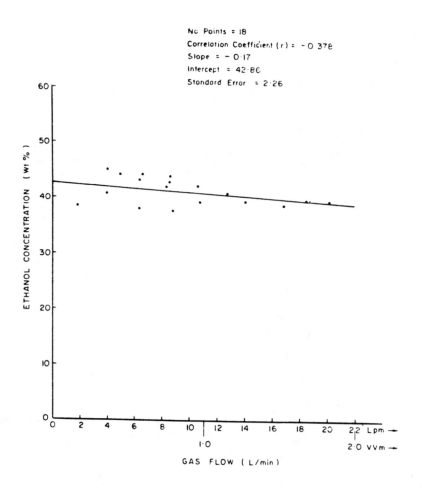

FIGURE 5.   ETHANOL CONCENTRATION IN THE VAPOUR PHASE
            vs GAS FLOW

240

The Parameter, σ, the specific volatile uptake, can, therefore, be calculated where $\sigma = r/f$, r is the collection rate of the volatiles and f is the gas flowrate.

EFFECTS OF NATURE OF SPARGING GAS

No changes were observed in vapour composition of volatilisation or condensation rates when the sparging gas was changed. Gases tested were air, helium, carbon dioxide and nitrogen. This surprising result is a reflection of near ideal behaviour in the gas phase of the volatile components under the operating conditions of low temperature and pressures (see below).

EFFECTS OF ETHANOL CONCENTRATION IN THE LIQUID PHASE

The vapour-liquid curve generated at 40°C in this work was very similar to that obtained at the boiling point and recorded in the literature, Figure 6 c.f. Figure 2. It was, therefore, concluded that

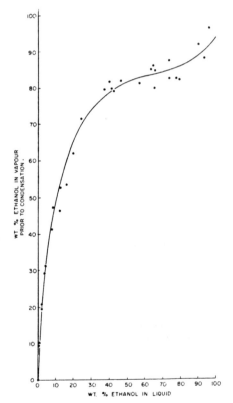

FIGURE 6.   VAPOUR - LIQUID CURVE (THIS WORK)

the ratio of ethanol to water in the vapour phase depended only on the
concentration in the liquid phase over the temperature range of
observation (40°-100°C). The influence of ethanol concentration on
the rate of evaporation of the volatiles will be dealt with in a later
section.

THE RATE OF EVAPORATION AND CONDENSATION

A mathematical relation has been derived which describes the
rate of volatile uptake with varying concentration. This is based on a
modification of the absolute humidity expression which is derived from
the gas laws, the only assumption being that ideality is approximated in
the vapour phase. The derived expression is

$$\sigma = \frac{1.4 \ M_{av} \Sigma fi}{M_g \ (P - \Sigma fi)}$$

where σ is the specific volatile uptake; 1.4 is a constant for the gas
used--here it is the density of moist air; M_{av} is the "averaged"
molecular weight, i.e. $M_{av} = \Sigma x_i MW_i$, where x_i and MW_i are
the mole fraction and molecular weight of the component i of the
mixture respectively. Σfi is the sum of the fugacities of the
components where fi, the fugacity of the i-th component, is
calculated from $fi = x_i \sigma_i P_i^{\circ}$ where x_i is the mole fraction
of the i-th component in the liquid, σ_i is the activity
coefficient of the i-th component obtained independently,
and, P_i° is the vapour pressure of the pure component i at the same
conditions of temperature and pressure.

M_g is the molecular weight of the stripping gas and P is the
total system pressure (usually 1 atmosphere).

Mass transfer rates from the liquid to the vapour phase are
obtained from

$$r = f\sigma$$

Where r is the volatilisation rate and f is the gas flowrate.

Excellent agreement was found between calculated and
experimental results, Figure 7. The apparent deviation at high ethaonl
concentrations can be attributed to poor activity coefficient data for
water at very low concentrations.

THE RATE AS A FUNCTION OF TEMPERATURE

Experimentally, it was found that the specific volatile uptake
approximately doubled for each 10°C rise in temperature in the range
40-60°C. Results calculated using the vapour pressures of pure ethanol
and water at 50° and 60°C instead of 40°C gave values closely in
agreement with experimental results, Figures 8 and 9.

242

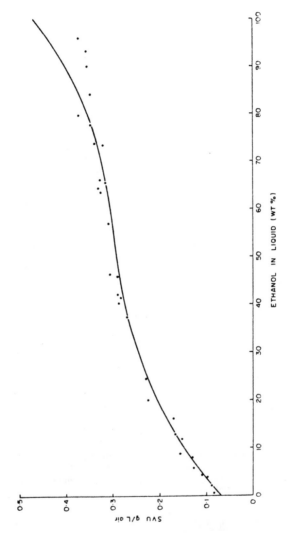

FIGURE 7.   PLOT OF SPECIFIC VOLATILE UPTAKE AS A FUNCTION OF ETHANOL
CONCENTRATION AT 40°C.
THE LINE IS FROM CALCULATION, THE POINTS ARE EXPERIMENTAL

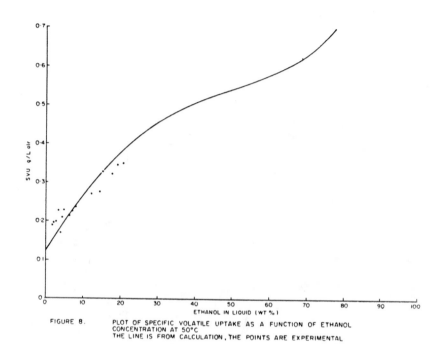

FIGURE 8.   PLOT OF SPECIFIC VOLATILE UPTAKE AS A FUNCTION OF ETHANOL
CONCENTRATION AT 50°C
THE LINE IS FROM CALCULATION, THE POINTS ARE EXPERIMENTAL

CONCLUSIONS

On a laboratory scale, gas stripping of dilute ethanol solutions
was found to be a significant option for the pre-concentration of ethanol,
free of cells and nutrients, thus reducing the final distillation costs
and permitting alternative separation technologies to be used. The equilibrium has been established and kinetic data obtained. An expression
has been derived by which prediction of volatile uptakes and condensation
efficiencies can be calculated over the temperature range of interest
(40-60°C).

The technique holds considerable promise for the continuous
stripping of continuous fermentations and work to evaluate this potential
is now under way. Results of this work will be reported elsewhere.

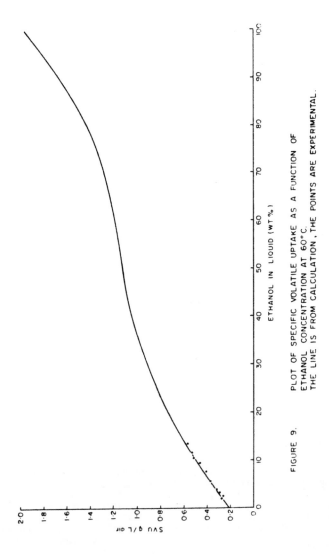

FIGURE 9.　PLOT OF SPECIFIC VOLATILE UPTAKE AS A FUNCTION OF ETHANOL CONCENTRATION AT 60°C. THE LINE IS FROM CALCULATION, THE POINTS ARE EXPERIMENTAL.

245

REFERENCES

1. Black, C., Distillation Modelling of Ethanol Recovery and Dehydration Processes for Ethanol and Gasohol CEP, September, 1980, 78-85.

2. Alcohol Fuels Program Technical Review, Solar Energy Research Institute, Winter, 1981.

3. Morris, E.C., Food Eng., Feb., 1981, p 93.

4. Hone, R.W., Lamarchand, M., and Malaty, W., Separation of Water-Ethanol Mixtures by Sorption, Part 2, Oak Ridge National Laboratory, Feb. 1981.

5. (a) Scheibel, E.G., Ind. Eng. Chem. $\underline{42}$ 1497 (1950).
   (b) Roddy, J. W., Ind. Eng. Chem. Proc. Res. Dev. $\underline{20}$ 104 (1981).
   (c) Leeper, S. A., Wankat, P.C., ibid $\underline{21}$ 331 (1982).
   (d) Roth, E.R., US Patent No. 4, 306, 884, December, 1981.

6. Eakin, D.E., et al, Preliminary Evaluation of Alternative Ethanol/Water Separation Processes, Batelle Memorial Institute, May, 1981.

7. Sheppard, J.D., and Margaritis, Biotech Bioeng. $\underline{23}$ 2117 (1981).

8. Cornell, L.W., and Montonna, R.E., Ind. Eng. Chem. $\underline{25}$ 1331 (1933).

RAPID PRODUCTION OF HIGH CONCENTRATIONS OF ETHANOL USING
UNMODIFIED INDUSTRIAL YEAST

W.M. Ingledew
Applied Microbiology and Food Science
University of Saskatchewan
Saskatoon, SASK. S7N 0W0

G.P. Casey
Department of Physiology, Carlsberg Laboratory
GL Carlsberg VEJ 10
DK 2500
Copenhagen, Valby DENMARK

Very high gravity brewing fermentations above 20° Plato (or Brix) stick or become exceedingly sluggish when syrups or sugars are used as the additional adjunct source. Yeast viability problems early in such very high gravity fermentations can be avoided by using 1 to 2×10^{6} CFU/ml per $^{\circ}$ Plato of extract. Even then, supplements of oxygen (or ergosterol and unsaturated fatty acid) and yeast-utilizable nitrogen are required to promote full attenuation. When oxygen and high levels of useable nitrogen are provided, rapid catabolism at 14°C occurs; up to 16% (v/v) ethanol can be made in 5-6 days in batch fermentations.

This conversion is mediated by unmodified production lager yeast and major losses in viability do not occur under these conditions. These results have led to a reassessment of ethanol tolerance in yeasts used in the beverage industry, and to an investigation of the potential uses of very high gravity "worts" in a number of fermentation industries.

INTRODUCTION

The alcoholic fermentation industry has evolved over the period of several thousand years. During that time, each sub-industry (brewing, wine, sake, and distilling) has developed its own unique manner of "wort" preparation and its own fermentation conditions. Thus for example, brewer's wort is clarified prior to fermentation, while distiller's wort is not. Likewise, brewery fermentations are batch fermentations, while sake fermentations involve the stepwise addition and saccharification of substrate. The starting gravity of fermentations in these industries also differs. Brewers traditionally ferment $11-12^{\circ}$ Plato worts (only 80% fermentable) but many have now increased to $16-18^{\circ}$ worts because of increased plant efficiency; reduced energy, labour, and capital costs; use of higher adjunct ratios; improved beer smoothness, flavour, and haze stability; and increased ethanol yields per unit of fermentable carbohydrate (9,17). Vinifications normally are initiated between 20-26° Brix, and alcohol destined for distillation is usually made from media containing 15 to 25% sugar (13). The cost per gal/day of fuel and feedstock alcohol production has been claimed to increase over a substrate level of 10% - primarily due to ethanol limitations on productivity (7).

The maximum level of ethanol capable of being produced in each of these fermentations is dependant in part on the level of useable sugars. Sugar concentrations however have been set because of the intrinsic ability of the different strains of **Saccharomyces** used within each industry to tolerate ethanol, and following experience with the range of fermentation problems (ie sluggish or stuck fermentations, and insufficient yeast crops) experienced. Ethanol toxicity and high osmotic pressure have been implicated as the limiting factors (8,14,16). Today, many believe that sake yeasts are more ethanol tolerant because they can produce "wines" of 20% (v/v), whereas brewers' yeasts make only 8-9% ethanol. The traditional and widely held order of ethanol tolerance in **Saccharomyces** yeasts is sake yeasts > wine yeasts > distillers' yeasts > brewers' yeasts.

The objective of our studies has been to work within the framework of existing industries - for example in brewing (4,5,6) to improve production economics by increasing the gravity of the original worts without altering significantly the final product, and in enology (Ingledew and Kunkee. J. Amer. Soc. Enol. Vitic. in press) by eliminating problems of sluggish vinifications and decreasing the time required for end-fermentation.

Previous work in this laboratory showed that useable nitrogen in a 12° Plato wort fermentation was deficient by the third day (10). Most brewers make high gravity worts using purified syrups to increase gravity (9). These syrups are very low in nitrogen (11), and their use compounds such nitrogen deficiency problems. The increases in gravity would in addition reduce normal levels of oxygen solubility (3). Oxygen is now known to be required for yeast synthesis of required sterols and unsaturated fatty acids (1) - it is in growth-limiting levels in wort, and brewers have learned to "rouse" or aerate wort prior to inoculation. It is also known that although the old belief in brewing was that the bulk of wort carbohydrate was converted to alcohol by largely non-growing cells, carbohydrate utilization by growing cells is far faster [the most rapid attenuation is during the period of new cell mass production] (12). It therefore seemed likely that in high gravity situations where extra carbohydrate must be metabolized, an increase in both the length and the level of cell mass would be required - ESPECIALLY IF A RAPID FERMENTATION WAS DESIRED. As we did not believe brewers' yeasts should be inhibited by 7-9% ethanol, we felt that brewers worts must be nutritionally inadequate in terms of nitrogen and oxygen - both of which would have to be increased at least proportional to increases in specific gravity.

Experiments were conducted in Wheaton Celstir bioreactors (Wheaton Scientific, Millville, N.J.) (Fig. 1). For many experiments, the headspace was flushed with O_2-free nitrogen to

Fig. 1.
Wheaton
Celstir
Culture
Flask

simulate as much as possible conditions in industry where produced CO_2 quickly purges the tank headspace (flasks have high surface to headspace volume ratios).

Experiments were carried out using a corn adjunct wort of 11.5 ^{O} Plato; - brewer's syrup was added to make very high gravity worts up to 34° P or more. Unmodified production strains of **S. carlsbergensis (uvarum)** (yeast slurry from a local brewery) were used. It was first found that a puzzling and unexpected immediate early loss in cell viability (up to 70%) took place during the first 12 hours after pitching (4). Such losses were minimized by high inoculation rates. Important to note here is that 6×10^6 CFU/ml is the traditional pitching rate used in breweries; 15×10^6 is the rate many breweries now use for 12° Plato and higher worts. In our experiments, much higher inoculation levels near 6×10^7/ml were needed to eliminate this early cell death (4). Methylene blue viable cell staining showed us parallel results and led us to question whether cells unable to form colonies on agar were also metabolically dead - i.e. could non viable cells still metabolize sugar to ethanol and CO_2.

To show this, cells were harvested from their fermentations at appropriate times, washed in buffer and CO_2 evolution was measured in a Warburg respirometer. Data for 12 hour cells is shown in Fig. 2 and is recorded in μL CO_2/mg dry weight yeast (in N_2 atmosphere). Note that in virtually all gravities of wort, CO_2 evolution was close to normal when the highest pitching rate was used. To cap off this part of the story, when $QO_2^{N_2}$ values (μL CO_2/hr/mg dry weight under N_2 gassing) were plotted over a range of total solids, the improved fermentation properties of yeast at a higher pitching rate were clearly seen (4). This could not be improved by adaption of yeast over 5 fermentation cycles. At a particular gravity, the fermentative vigor of the yeast pitched at a higher pitching rate ON A PER MG BASIS is superior to yeast pitched at a lower rate.

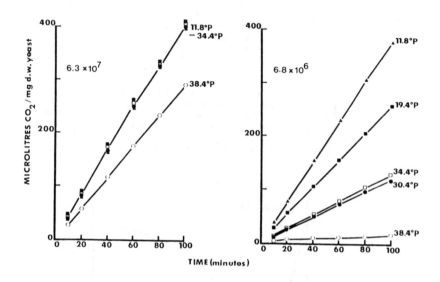

Fig. 2. CO_2 Production from Yeasts Harvested at 24 Hours from 11.8 to $38.4°$ Plato Worts

The cause of cell death is unknown but it was not ethanol; no detectable ethanol was made in the first 12 hours of fermentation. Osmotic pressure, shock excretion or severe dilution of intracellular nutrients are suspected.

We therefore can summarize with the following (4):
1. As wort original gravity increases, early losses in yeast viability occur, with increasing severity.
2. Such losses in viability are reflected by losses in fermentative ability as well.
3. Severity of these losses can be significantly reduced by the use of higher pitching rates.
4. High pitching rates should therefore be considered for fermenting high gravity worts.

Although we were not the first to suggest higher pitching rates in high gravity brewing - this data is proof of a largely empirical finding.

To bring these findings into perspective with the literature and with the rest of this investigation it should be recalled that Kirsop has shown in $12°$ P wort that the efficiency with which dissolved O_2 is used for lipid synthesis <u>decreases</u> with <u>decreasing</u> pitching rate (2). Furthermore, only 30% of O_2 is

250

used by yeast for sterols and fatty acids. O_2 utilization continues after sterol synthesis is complete. The quantity incorporated into sterols is therefore proportional to inoculum size especially in high gravity worts. For this reason and because of the above viability problems, we used higher than traditional pitching rates to determine the factors required for full attenuation of very high gravity worts. However, it is important to note that prevention of sticking fermentations could not be overcome solely by the proposed increase in pitching rate.

In planning further experiments we kept in mind our prior observations that the brewing fermentation is deficient in useable N and that O_2 may also be growth limiting. Increased growth was considered important due to Kirsop's findings that carbohydrate utilization in high gravity worts occurs faster and for a longer duration when cells were still actively growing.

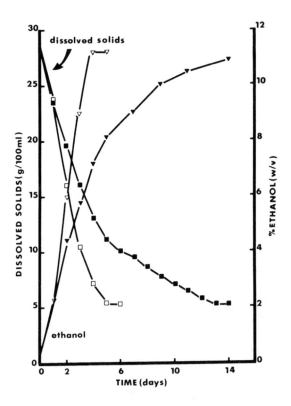

Fig. 3 Fermentation of 28% Wort with (open) and without (closed) Headspace Nitrogen Flushing.

In a simple experiment (Fig. 3.) it was possible to show the possible role and need of O_2 in 28° P worts. These fermentations were carried out in 1 L Celstir flasks with and without N_2 gas flusing of headspace. Flasks without flushing were plugged in the side ports with foam plugs, permitting limited but important air access. Nitrogen flushed flasks "stuck" with high levels of sugar unused. Flasks with air access fermented to completion in 12 days. Cell synthesis was dramatically increased by air access.

The above partly aerated fermentation, although successful, was however rather lengthy so attempts were made to enhance it by nutritional supplementation. (Table I). Very high gravity worts were made by using 11.5° P corn adjunct wort to which brewing syrup was added to 27° P. Additional nutrient supplements were added as follows: a) unsupplemented; b) 1% yeast extract (YE); c) 40 ppm ergosterol + 0.4% Tween 80 (ERG, TWEEN); d) YE, ergosterol and Tween 80.

TABLE I

FERMENTATION OF N_2 FLUSHED 27° P WORT

	NET CELL SYNTHESIS mg/mL (day)	FAN LEVELS mg/L orig. end used			MALTOSE g/100
UNSUPPLEMENTED	5.6 (3)	213	107	106	2.65
YEAST EXTRACT	6.2 (3)	616	509	107	2.24
ERG & TWEEN	9.8 (4)	213	48	165	0.92
YE, ERG & TWEEN	10.2 (4)	633	344	289	0.85

Ergosterol and oleic acid (present at 0.6% in Tween 80) have been shown by Thomas and Rose (15) to be deficient in anaerobically grown yeast. Increases in these compounds (via nutrient supplementation) resulted in yeasts that were more tolerant to ethanol in **in vitro** experiments. Oxygen is required by yeasts for **de novo** synthesis of ergosterol and oleic; anaerobically, reproductive growth ceases when a limiting value of sterols is reached.

There appeared to be a synergism between yeast extract and ergosterol-Tween with respect to free amino nitrogen (FAN) utilization (Table 1). Ergosterol and Tween by themselves stimulated most of the cell mass increase and promoted the maltose utilization. Normal profiles (similar to unsupplemented fermentations) were seen when yeast extract was present alone in these anaerobic fermentations. Fig. 4 shows the progress of these fermentations vs. time.

Interestingly the same experiments done with air access (semi anaerobic) demonstrate no need for lipids - supplementations with

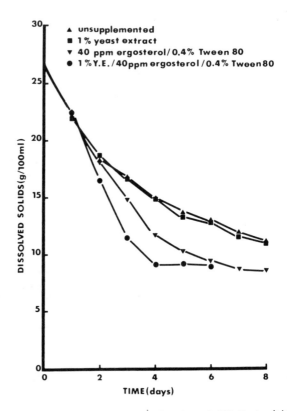

Fig. 4 Fermentation under N$_2$ Gassing of 27% Wort with and without Supplementation.

yeast extract alone result in rapid rates of fermentation and increased cell mass synthesis (6). In this case some ergosterol and oleic are made by the yeast from O$_2$ in the flask headspace. Supplementations are still effective when added later in fermentation (5) but not if added when growth has completed. This certainly helps to prove that the sticking was not due to ethanol toxicity but rather to nutritional deficiencies.

We also have conducted experiments with brewers wort to show that sparging with O$_2$ (100 mL O$_2$/min/L) for 4 hour periods in the first two days of fermentation will substitute for the need for preformed lipids. Assimilable nitrogen (in this case FAN) must however be increased to get full stimulation. This probably also indicates that sticking here is not due to EtOH toxicity — but instead is due to nutritional deficiency (or at least the need for membrane repair).

253

The levels of supplementation have now been optimized for fermentation of 27° Plato wort, as we have found that 0.8% yeast extract and 24 ppm ergosterol with 0.24% Tween 80 promote full and rapid attenuation. Further reductions of these supplements led to prolongation of the fermentation time.

Under these conditions in 27-30° Plato brewers wort, end fermentation occurs with 8-10% total solids remaining. This is because wort sugars from malt and from corn syrup contain about 30% dextrins which cannot be totally hydrolyzed to fermentable sugar by α-amylases and β-amylases. We have also conducted experiments with added maltose instead of syrup. Maltose brings the gravity to near 30° Plato from the 12° P provided by the corn adjunct commercial wort. In this way, a more complete attenuation is experienced, and final ethanol levels over 14% v/v can be obtained. This is not bad for a yeast which in the literature is ascribed to a tolerance of only 7-9%. This fermentation is shown in Figure 5 with and without supplementation.

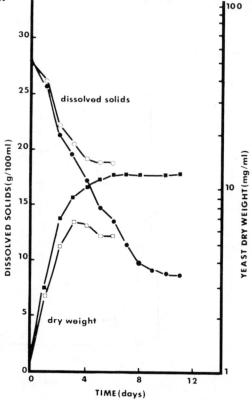

Fig. 5 Fermentation of 28% Maltose-Adjunct Wort with (open) and without (closed) Full Supplementation.

Again, we believe that protracted fermentation in the unsupplemented fermentation is not due to EtOH toxicity but to
slower rate of carbohydrate utilization by nutrient starved,
stationary phase cultures of lesser dry weight/mL.

Repitching experiments with yeasts from these very high
gravity fermentations carried out over 5 generations demonstrated
that: 1) no deterioration in rate and extent of fermentation
occurred with supplemented yeast; and 2) in unsupplemented
fermentations one sees a decrease in rate of substrate
utilization, a decrease in extent of attenuation, and a
corresponding increase in time required. This is best visualized
by comparing Run 1 with Run 5 (supplemented and unsupplemented,
Fig. 6.).

Supplementation has now been refined as more information has
come available. For example, both ergosterol and unsaturated
fatty acid are required for full supplementation. Oleic alone is
more stimulatory than ergosterol alone. Lipids in yeast
membranes increase from 0.65 to at least 0.85% when
supplementation is practiced, and end fermentation lipid levels

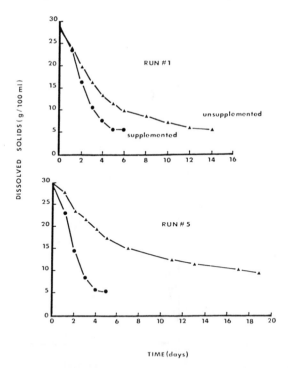

Fig. 6. Comparison of First to Fifth Successive High Gravity Wort
Fermentation with and without Supplementation.

are higher in supplemented yeast than in unsupplemented yeasts (0.62 vs 0.48). Yeast extract was used for most of the described experiments.

Preliminary work has also indicated that trypticase, casamino acids and inorganic ammonium ion or urea will satisfy yeast requirements. A high malt wort (no corn grits) at 12° Plato with syrup to supply 30% dissolved solids was found to need only 0.2% YE for full nitrogen supplementation. The nitrogen supplementation for rapid fermentation of wine has also been extensively examined (Ingledew and Kunkee, in press). A maximum ethanol level in high maltose wort approaches 16.2% v/v, prescribing a limit of approximately 32% dissolved solids.

Figure 7 illustrates the important but often disregarded relationship between temperature, rate of fermentation and cell viability. Certainly, as temperature increases, the speed of fermentation increases (14°C to 30°C). Final alcohol levels are similar in each case. End viability, however decreases markedly at higher teperatures - i.e. only 0.01% cells will survive at 30°C, 39% at 25°C, and 85% at 20°C. At

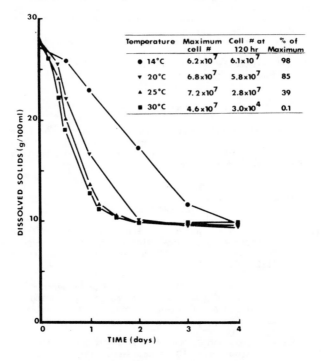

Temperature	Maximum cell #	Cell # at 120 hr	% of Maximum
● 14°C	6.2×10^7	6.1×10^7	98
▼ 20°C	6.8×10^7	5.8×10^7	85
▲ 25°C	7.2×10^7	2.8×10^7	39
■ 30°C	4.6×10^7	3.0×10^4	0.1

Fig. 7. Fermentation of Fully Supplemented Worts at 14°C, 20°C, 25°C, and 30°C.

14^OC, virtually the entire population is viable at end fermentation. It is therefore certain that one must observe care in designing fermentation parameters if: repitching is required, if autolytic off-flavours are detrimental to the final product, or if ester levels are desired to be low.

In summary, (5,6) we have been able to demonstrate that:

1) Unmodified commercial brewing yeast under nutritional supplementation and increased pitching rate can, in 1 week at 14 ^{O}C, produce and tolerate up to 16% v/v alcohol. No loss in viability is evident at these temperatures.

2) Earlier reports indicating that $16-18^O$ P limits in high gravity brewing are due to slow fermentation due to ethanol toxicity are probably false.

3) For fermentation of high sugar levels, adequate elevated yeast levels and prolonged growth are required. The latter conditions can be met by 0_2 or nutritional substitution for 0_2, and by increasing FAN levels with a suitable N source.

This work should provide incentive towards continued studies on high gravity wort fermentation, advanced studies on tolerance to alcohol in the genus **Saccharomyces,** understanding of the factors necessary for production of high levels of alcohol for fuels and feedstock, and the understanding of stuck and sluggish fermentations in industry.

ACKNOWLEDGEMENTS

The Natural Sciences and Engineering Research Council and Molson Breweries of Canada Ltd. are thanked for provision of research grants which have fostered basic research into practical problems of relevance to industry.

LITERATURE CITED

1. Andreasen, A.A., and T.J. Stier. 1953. Anaerobic nutrution of **Saccharomyces cerevisiae** I. Ergosterol requirements for growth in a defined medium. J. Cell. Comp. Physiol. 41:23-36.

2. Aries, V., and B.H. Kirsop. 1977. Sterol synthesis in relation to growth and fermentation by brewing yeast inoculated at different concentrations. J. Inst. Brew. 83:220-223.

3. Baker, C.D., and S. Morton. 1977. Oxygen levels in air-saturated worts. J. Inst. Brew. 83:348-349.

257

4. Casey, G.P., and W.M. Ingledew. 1983. High gravity brewing: Influence of pitching rate and wort gravity on early yeast viability. J. Amer. Soc. Brew. Chem. 41:148-152.

5. Casey, G.P., C.A. Magnus, and W.M. Ingledew. 1983. High gravity brewing: Nutrient enhanced production of high concentrations of ethanol by brewing yeasts. Biotech. Lett. 5:429-434.

6. Casey, G.P., C.A. Magnus, and W.M. Ingledew. 1984. High gravity brewing: Effects of nutrition on yeast composition, fermentative ability, and alcohol production. Appl. Environ. Microbiol. 48:639-646.

7. Cysewski, G.R., and C.R. Wilke. 1976. Utilization of cellulostic materials through enzymatic hydrolysis. I. Fermentation of hydrolysate to ethanol and single cell protein. Biotech. Bioeng. 18:1297-1313.

8. Day, A., E. Anderson, and P.A. Martin. 1975. Ethanol tolerance of brewing yeasts. Eur. Brew. Conv. Proc. 15:377-391.

9. Hackstaff, B.W. 1978. Various aspects of high gravity brewing. M.B.A.A. Tech. Quart. 15:1-7.

10. Ingledew, W.M. 1975. Utilization of wort carbohydrates and nitrogen by **Saccharomyces carlsbergensis**. M.B.A.A. Tech. Quart. 12:146-150.

11. Jones, M., and J.S. Pierce. 1964. Some factors influencing the individual amino acid composition of wort. J. Amer. Soc. Brew. Chem. 22:130-136.

12. Kirsop, B.H. 1982. Developments in beer fermentation. Topics in Enzyme and Ferm. Biotechnol. 6:79-131.

13. Lyons, T.P. 1981. A Step to Energy Independence. Alltech Inc. Lexington, Kentucky.

14. Owades, J.L. 1981. The role of osmotic pressure in high and low gravity fermentations. M.B.A.A. Tech. Quart. 18:163-165.

15. White, F.H. 1978. Ethanol tolerance of brewing yeast. Inst. Brew. (Australia and New Zealand Sect.) Proc. Conv. 15:133-146.

16. Whitear, A.L. and D. Crabb. 1977. High gravity brewing - concepts and economics. The Brewer 63:60-63.

BIOPROCESS MODELLING AND OPTIMIZATION USING ACETONE-BUTANOL SYSTEM

B. Volesky - Biochemical Engineering Unit
McGill University, Montreal, Québec H3A 2A7

Abstract

The acetone-butanol fermentation process has been selected as a model process to develop and demonstrate the use of the mathematical model and computer simulation of bioreactor performance. Methodology of the model development is presented together with the set of equations representing models for the batch and continuous-flow culture systems, respectively. Selected examples of steady-state analysis of bioreactor behaviour for two continuous-culture arrangements show existence of multiple steady states. The potential of the bioprocess model in guiding experimental work and in process optimization by on-line dynamic control is pointed out.

INTRODUCTION

Mathematical modelling of biological systems, processes and cellular metabolic activities has entered a new phase. While the first-generation modelling attempts were based on over-simplifying assumptions and led to often unrealistic conclusions based on crude and simple models, the currently advancing second-generation models are based on advanced knowledge of process kinetics, hydraulics and mass transfer aspects supported by closing of material, energy and electron balance equations. Development of powerful and easily available computer hardware coupled with recently developed highly sophisticated software means enables solution of even very "stiff" sets of differential equations representing the second-generation process modelling attempts. This new generation of models presents us with a very powerful tool which, according to the most recent experience, can be used in:

1) guiding laboratory experimentation, drastically reducing the volume of essential experimental work;
2) on-line process control and optimization;
3) revealing and subsequently avoiding metabolic and mass transfer problems on a cellular or even molecular level.

The process of microbial conversion of carbohydrate-type materials into the mixture of butanol-acetone-ethanol has been selected as a model process to be studied in some detail for several reasons. Although all of the solvents produced by this fermentation are of undisputable industrial interest, the economics of this biosynthetic process still does not make it currently competitive enough with hydrocarbon-based petrochemical processes. However, its intriguing nature and aspects which are common to many microbial processes currently employed or with a high commercial potential, make the study of the acetone-butanol (A-B-E) process a challenging and relevant task.

Butanol biosynthesis from carbohydrates is a strictly anae-robic process catalyzed by certain Clostridium bacteria. Small amounts of acetone and ethanol by-products result in the fermenta-tion which proceeds via acetic and butyric acids metabolic interme-diates, normally accummulating during the initial growth period of the standard batch culture[1],[2]. In the growth non-associated product formation period, the intermediates are subsequently converted into the solvents. The solvents typically never exceed total concentra-tion of 2% in the broth representing the inhibitory threshold. Inhibition of the bioconversion represents a major drawback of the conventional mixed-vessel batch system which has a definite limita-tion based on the inhibited batch kinetics. A great deal of con-troversy surrounds also the culture physiology[3] which could not be properly investigated in an inherently unsteady-state batch system. Unconventional fermentation techniques based on the continuous-flow culture method can be employed to minimize the inhibitory effect of the accumulating toxic end-product(s) and a steady state regime in the bioreactor can be maintained. Modified bioreactor design can enable achievement of high cell (biocatalyst) densities resulting in increased bioreactor productivities and improved process eco-nomics. Advancement of the immobilized cell technology offers a new "lease on life" particularly for sequential biosynthetic pro-cesses such as the acetone-butanol fermentation because of the different type of kinetics of fixed-bed or similar (bio-)reactors which is infinitely better suited to the nature of the process. The flow-through continuous packed-bed reactor arrangement is not only superior arrangement for achieving much higher productivities but it also offers a new powerful tool for studying the process under steady-state conditions not attainable otherwise.

Experiments concerning the kinetics of fermentation processes are mostly carried out with two types of culture systems, namely in a batch or a continuous-flow arrangement. The batch system is experimentally simpler and the data collection is faster than in the continuous-flow mode. On the other hand, the mathematical model of the batch cultivation leads to a system of ordinary dif-ferential equations of the first order with initial conditions while the one for the continuous culture system results in a set of non-linear equations. Solution of a system of non-linear equations has become routine.

METHODOLOGY

Formulation of a mathematical process model, together with an adequate description of the key physiological factors, is the first step toward simulation of a biological process. The mathematical model of the process is determined by its coefficients (parameters) which have to be determined on the basis of experimental data.

This procedure of model "identification" may be relatively the most difficult part of the process systems analysis because it in-volves model fitting procedures.

The experimentally determined integral functions for the de-pendent variables are thereby fitted to the corresponding differ-ential mass balances usually written in the form of a system of nonlinear first-order ordinary differential equations with given initial condtions. From the methodological viewpoint, it is appropriate to divide the mathematical model identification procedure into three phases. The first phase, consisting of expressing values of rates of individual processes, is followed by assigning appropriate types of functional descriptions to the data. Finally, the "model fitting" to the data points is carried out.

The first phase of the model identification procedure can be automated by using a computer system equipped with suitable software subroutines permitting smoothing of the experimental relationships. In general, the following relationship is dealt with:

$$y_i(t) = f_i(t) + \varepsilon_i$$

where y_i is the measured value,
   ε_i is the measuring error, and
   f_i is the approximating function

The measurements are done in a time series $t_1; t_2; \ldots t_N$ with a corresponding series of experimental values $y_{i_1}; y_{i_2}; \ldots y_{i_N}$. Estimate of the measuring error ε_i can be independently done based on several known formulae. In case of the fermentation process considered, the estimate of this error value is in the range 0.1-0.5 of the absolute measured value for each independent variable. For smoothing of the experimental data it is possible to use some of the approximation methods available in the literature. However, for the best curve fitting and for the most accurate estimate of the numerical value for the dependent variable differential y_i with time, it is better to use a piecewise polynomial approximation method based on the theory of so called "spline functions". Details of this approach can be found, for example, in the work of deBoor[4]. While the methodology to be used here for evaluation of specific velocities of individual fermentation reactions is also based on the above work, the approximation of experimental data uses a complex of programs (SMOOTH[4]) to minimize the following function over all the values of the approximating polynomial and its m-th derivative:

$$p \sum_{i=1}^{N} \left[\frac{y_i - f(t_i)}{\varepsilon_i} \right]^2 + (1-p) \int_{t_1}^{t_N} [f^m(\tau)]^2 \, d\tau$$

This approach to minimizing the above relationship represents a compromise between the requirement for the approximation curve to be the closest to the experimental data on one hand and for the relationship to be the smoothest on the other. This choice is determined by the magnitude of ε_i estimate which, in turn, automatically determines the value of weight coefficient p. Again, further details and description of the algorithm is available in the abovementioned work of deBoor[4] (Chapter XIV). The computer program published in his book was transcribed from computer language FORTRAN 77 into FORTRAN IV and has been de-bugged in the interactive computer system of McGill University (MUSIC).

For the experimental data analysis, programs SMOOTH, CHOLID and SETUPQ will be used, supplemented by the main program enabling input of data. Simultaneous transformation and smoothing of the results will be facilitated through the use of subroutine TRANSF.

The smoothed values of dependent variables and their first derivatives will be used in this program according to the users demand for calculations concerning specific individual reactions in the fermentation process. Particularly important for the subsequent process model formulation are the culture specific growth rate μ, the substrate utilization and the product synthesis rates r_s and r_p, respectively. For the batch process, these rates are defined by the following relationships:

$$\mu = (1/X)(dX/dt) \qquad r_s = (1/X)(dX/dt) \qquad r_p = (1/X)(dX/dt)$$

For continuous-flow cultivations, modified relationships for r_s and r_p have to be used:

$$r_s = (1/X)[D(S_0-S)-(dS/dt)] \qquad r_p = (1/X)[(dP/dt)-DP]$$

The second phase of the process modelling follows in which a biologically interpretable functional relationship could be assigned to the numerical values of corresponding rates. However, it is first necessary to specify the limitations of the relationships considered here, together with simplifying assumptions implicitly included in the model.

RESULTS

The model for a batch culture was derived for experimental conditions where:

1) there is no process limitation by nitrogen source
2) functional relationships are valid for concentration ranges:
   glucose 0-50 g/L, biomass 0.03-10 g/L
   pH 5-6, and temperature of 38°C,
3) product concentrations would not exceed limits:
   acetone 5 g/L, ethanol 1.5 g/L, butanol 0-11 g/L,
   organic acids 5 g/L,
4) data were derived from batch cultivations
   (and from a filtered continuous culture at $D = 0.113$ h[-1]).

For derivation of the kinetic relationships, the information on bio-chemical pathways involved is to be used which has been summarized by Volesky elsewhere[1,2,5]. From this biochemical scheme, it is apparent that butanol acetone, ethanol, biomass, CO_2 and H_2 are the end-products of the metabolism while butyric and acetic acids are intermediate compounds connected with the growth kinetics. As a matter of principles, conclusions have to be incorporated in the process model.

In the mathematical formulation of the anaerobic process dynamics the primary role is played by the cell growth kinetics reflecting the cell physiology during the start-up and culture production production phase. To describe the direction of the physiological culture activity, Powell[6] introduced the term "metabolic activity functional" Q defined in terms of the specific growth rate. He recommends to relate the latter metabolic functional to a suitable marker of physiological state which can be biochemically determined. The concentration of RNA can be used which is linearly related to growth.

The relationship $\mu(S,B)$ is determined as a characteristic of culture and it could be expressed from the experimental data. For mathematical description of this relationship for the acetone-butanol process, it is possible to choose a kinetic formula based on simultaneous limitation by the substrate and the product inhibition established in the literature as Yerusalimski-Monod equation. Since the value of saturation constant is rather high for growth, it is possible to use a linear relationship with respect to substrate resulting in the following function:

$$\mu(S,B) = K_1 \, S \, \frac{K_I}{K_I + B}$$

The biomass grown during the process is eventually chemically attacked by the product synthesized with resulting poisoning and lysis of the cells[3]. This phenomenon can be modelled by simple kinetics of the first order with regard to the biomass and butanol. The differential mass balance for the biomass (X) is then expressed. It can be used for modelling of the initial lag phase as well as the final cell lysis caused by the product.

The cyclical biosynthetic pathway for producing acetone, butanol, ethanol, and gaseous end-products is outlined in the general scheme on Figure 1. The sugar consumption corresponds to the

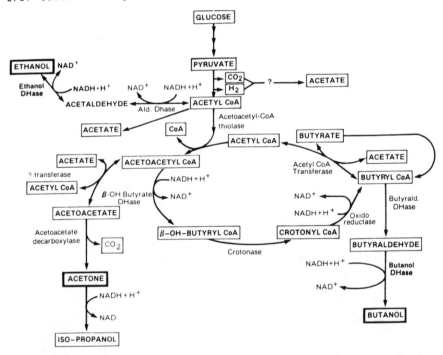

FIGURE 1. Biosynthetic pathway leading to the accumulation of metabolites in *Clostridium acetobutylicum*.

two main biochemical pathways operating in parallel. One pathway results in production of acetic acid, acetone, ethanol and gaseous products, all products of the anaerobic energetic metabolism. The other pathway produces cellular mass, butanol and butyric acid. According to this hypothesis concerning the biochemical pathways, the sugar consumption balance can be modelled by another differential equation. Since butanol and butyric acid are synthesized in the same part of the sugar-metabolizing system, there is a mutual interaction in their synthesis whereby increasing production of butyric acid results in decreased production of butanol and vice versa. This situation can be modelled as a competition for substrate. A conversion factor is required to correlate the butyric acid production rate with that for butanol through the ratio of molecular weights of the two compounds which is 0.831. The product formation is determined by the substrate feed into the metabolic system as could be expressed by:

$$k_7 S = r_B + 0.831\ r_{BA}$$

The dynamics of butanol synthesis can then be expressed by another differential mass balance and similarly for acetic acid which is produced mainly in the second branch of the metabolic scheme describing glucose utilization and its relationship to the acetone production being similar to that existing between butanol and butyric acid. The kinetics of acetone production is connected with the production of acetic acid in a similar way as the case is for butyric acid and butanol. Based on the acetone production and its dependence on the concentration of acetate sub-units in the metabolism, the dynamics of acetone biosynthesis can be described by another differential equation.

The kinetics of ethanol production is not affected by other metabolic products and the ethanol production rate can be described by a Monod-type function relating it to the substrate concentration. The kinetics of the gaseous product formation can be described in a similar manner. The dynamics of CO_2 and H_2 evolution is thus given by yet another set of differential relationships.

The system of differential equations summarized in Table 1 represents the mathematical model of a batch culture.

The A-B-E process kinetics model can be extended to cover the conditions of the continuous-flow culture. The system dynamics at a constant broth volume (V) and a dilution rate D= F/V can be described by the set of differential equations summarized in Table 2 which are based on the process species mass balances.

Determination of the magnitude of individual coefficients in the differential equations represents the third step in the identification of the mathematical process model. This task constitutes a somewhat different problem. It is done as a standard task in locating a numerical extreme. This approach is greatly facilitated by utilizing the computer.

Obviously, the model systems are derived from and based upon corresponding sets of original laboratory experimental data obtained from various operating process alternative arrangements.

TABLE 1

BATCH CULTURE MODEL

$$\frac{dy}{dt} = \mu \ (S,B) \ y + 0.56 \ (1-y)y$$

$$\mu (S,B) = k_1 \ S \ \frac{K_I}{K_I + B}$$

$$\frac{dX}{dt} = 0.56(y-1) \ X - k_2 \ XB$$

$$\frac{dS}{dt} = -(k_3 S + k_4 \ \frac{S}{S + K_S})X$$

$$\frac{dBA}{dt} = (k_5 S \ \frac{K_I}{K_I + B} - k_6 \ \frac{BA}{BA + 0.5})X$$

$$\frac{dB}{dt} = k_7 SX - 0.831 \ \frac{dBA}{dt}$$

$$\frac{dAA}{dt} = X(k_8 \ \frac{S}{S + K_S} \ \frac{K_I}{K_I + B} - k_9 \ \frac{AA}{AA + K_{AA}})$$

$$\frac{dA}{dt} = k_{10} \ \frac{S}{S + K_S} \ X - 0.498 \ \frac{dAA}{dt}$$

$$\frac{dE}{dt} = k_{11} \ \frac{S}{S + K_S} \ X$$

$$\frac{dCO_2}{dt} = k_{12} \ \frac{S}{S + K_S} \ X$$

$$\frac{dH_2}{dt} = k_{13} \ \frac{S}{S + K_S} \ X$$

NOMENCLATURE :

Concentrations

A... acetone
AA.. acetic acid
B... butanol
BA.. butyric acid
CO_2 carbon dioxide gas
E... ethanol
H_2.. hydrogen gas
S... substrate
X... biomass

265

TABLE 2

CONTINUOUS-FLOW CULTURE MODEL

$$\frac{dy}{dt} = \left[k_1 \frac{S}{S+K_S} \frac{K_I}{K_I+B} - 0.56(y-1) \right] y$$

$$\frac{dX}{dt} = 0.56(y-1)X - k_2 BX - DX$$

$$\frac{dS}{dt} = -k_3 SX - k_4 \frac{S}{S+K_S} X + D(S_o - S)$$

$$\frac{dBA}{dt} = k_5 \frac{S}{S+K_S} \frac{K_{IBA}}{K_{IBA}+B} X - DBA$$

$$\frac{dB}{dt} = k_7 \frac{S}{S+K_S} X - 0.831 \left(\frac{dBA}{dt} + D.BA \right) - DB$$

$$\frac{dAA}{dt} = k_8 \frac{S}{S+K_S} \frac{K_I}{K_I+B} X - D.AA$$

$$\frac{dA}{dt} = k_{10} \frac{S}{S+K_S} X - 0.5 \left(\frac{dAA}{dt} + D.AA \right) - D.A$$

$$\frac{dE}{dt} = k_{11} \frac{S}{S+K_S} X - DE$$

$k_1 - k_{13}$ .... model parameters

K_I ..... inhibition constant

K_S .....saturation constant

y .....dimensionless "marker of the physiological state" of the culture.

$$y = \frac{RNA}{RNA_{max}}$$

μ .....specific culture growth rate

The experimental data constituting the basis for model identification will not be reported in this work. Instead, the most interesting features and conclusions of computer process simulations will be briefly indicated.

It was of interest to analyze the steady-state performance of the biosystem in the area of high productivity of neutral solvents. Steady-state values of system parameters can be obtained by numerically solving the set of non-linear equations originating from the set of differential equations for the continuous-flow culture system whereby their left sides are equal to zero. The solution showed that for the continuous-flow there could be up to three steady states. In the very informative plot of broth butanol vs. sugar concentration, the high-butanol steady state appears more stable than the lower one (Figure 2). Exactly opposite is

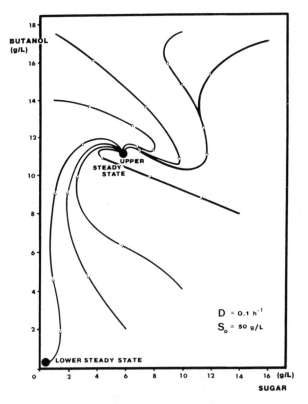

ANALYSIS OF STEADY STATE

FIGURE 2. Continuous-flow culture of *Clostridium acetobutylicum*. Bioreactor performance at a steady state as predicted by the system model.

the case with the cell-recycle (or retention) whereby the model computer simulation and steady state analysis indicated two steady states: one when the process operates under sugar limitation characterized by high biomass and lower butanol concentrations, and another one characterized by end-product inhibition, higher characterized by end-product inhibition, higher butanol and lower biomass concentrations. Further increase in the inlet sugar concentration results in the two steady states drawing closer together, eventually becoming one at 40 g/L of sugar in the feed. Increasing the sugar inlet concentration beyond this value amounts to destabilizing the bioreactor system which under these operating conditions could not maintain a steady state.

These indications derived from the computer process simulations based on the mathematical model are interesting indeed and can serve in devising the process control strategy for optimization of the bioreactor performance.

CONCLUSIONS
A brief review of selected main features of the work done on the modelling and analysis of the acetone-butanol fermentation has been attempted. Mathematical models were presented for the batch and continuous-flow operation of the bioreactor which reflect performance of the real system. Analysis of the bioreactor behaviour was done based on the computer simulation of its operation. It was demonstrated that the results of such model simulation can serve as a basis for appropriate design of experimental work as well as for devising a strategy for optimization of a process with multiple process parameters.

ACKNOWLEDGEMENT
This work was generously supported by National Research Council of Canada contracts as part of the Fermentation Technology and Bioenergy Development programs directed respectively by Doctor W.G.W. Kurz and Dr. R. Overend.

REFERENCES
1. Volesky, B., Mulchandani, A., and Williams, J. 1981. Biochemical production of industrial solvents from renewable resources. Annals NY Acad. Sci. 369, 205-218.
2. Volesky, B. 1982. Development of Fermentation Techniques for the Production of Industrial Solvents From Renewable Resources. Final - NRC/DSS Research Contract #07SZ.31155-8-6413. NRC Saskatoon, Saskatchewan.
3. Westhuizen, Van Der A., Jones, D.T. and Woods, D.R. 1982. Autolytic activity and butanol tolerance of Clostridium acetobutylicum. Appl. Envir. Microbiol., 44 (6-Dec.).
4. de Boor, C. 1978. A Practical Guide to Splines. Springer Verlag, N.Y.
5. Yerushalmi, L., Volesky, B., Leung, W.K. and Neufeld, R.J. 1983. Variations of solvent yield in the acetone-butanol fermentation. Eur. J. Appl. Microbiol. 18, 279-86.
6. Powell, E.Q. 1968. Transient changes in the growth rate of microorganisms. Proc. 4th Symp. on Cont. Cult. of Microorg. (Malek, I., ed.). Academia, Prague, 275-84.

HIGH RATE REACTORS FOR METHANE PRODUCTION

L. van den Berg
Division of Biological Sciences
National Research Council
Ottawa, Ontario, K1A 0R6

ABSTRACT

Fermentation systems for the production of methane from industrial waste waters have developed rapidly in the last 15 years and commercial use is rapidly expanding. The new systems of advanced reactors are based on retention of the methane producing bacteria, especially those converting acetic acid to methane. The methods of retention vary from flocculation and settling, to attachment to stationary support surfaces. The resulting increased concentration of microorganisms has increased the rate of methane production, has reduced costs and has drastically reduced problems of instability in anaerobic waste treatment. Canadian developments in this field, from laboratory studies to commercial application and including development of a new type of reactor, are emphasized.

INTRODUCTION

Industrialization and urbanization, with the concomitant serious waste disposal problems, resulted in the widespread practise of methane recovery from primary and activated sludge (van Brakel, 1980). Developments in the methanogenic process, however, were few and far between from the turn of the century to about 1950. It was recognized that temperature affected the process drastically and that mixing was beneficial. Temperature control (using the methane produced as a source of energy) and mixing were therefore introduced over the years. Process control improved with the observation that organic (volatile) acids accumulated under sub-optimum conditions and that the amount and the composition of the gas produced were important indicators of the health of reactors. Major improvements, however, depended on a more thorough understanding of the process. Unfortunately, the extreme anaerobic requirements of the bacteria involved proved a stumbling block until better methods for studying anaerobes became available after 1950 (Hungate, 1950).

Of major significance in the development of better anaerobic methanogenic reactors was the realization that much of the performance of anaerobic digesters depended on the conversion of acetic acid to methane and that the bacteria involved grew very slowly (mass doubling times of 8-80 days under practical conditions). This agreed with early empirical observations that hydraulic residence times in excess of 15 days were required for satisfactory operation. The newer types of reactors succeeded in extending the retention time of the bacteria in the reactor (to usually well over 15 days) while at the same time reducing hydraulic residence times (reactor volume divided by daily volumetric flow) to a few hours or a few days. The concentration of bacteria in the reactor was increased at the same time, leading to markedly increased rates of conversion of organic material to methane.

METHANE PRODUCTION

The rate of methane production of a reactor is proportional to the concentration of organic material in the substrate solution, and the fraction of this material that will be converted to methane. This rate is also inversely proportional to the hydraulic residence time and the oxygen content of the organic material converted to methane. For example, carbohydrates yield less methane than fats (0.3 and up to 1 m^3/kg respectively).

Many stirred tank reactors (municipal digesters for example) have volumetric methane production rates of only about 0.5 m^3/m^3/day with hydraulic residence times of over 20 days. The stoichiometry of the conversion reactions indicates that these rates of methane production are only possible with substrate concentrations of at least 35 kg/m^3. To treat more dilute wastes, even at these low methane production rates, it is necessary to reduce hydraulic residence times. As mentioned, this is possible in advanced type reactors in which active biomass is retained in high concentrations, in spite of short hydraulic residence times. In addition, these advanced reactor types can have methane production rates of over 5 m^3/m^3 reactor/day at hydraulic residence times of less than half a day.

CHARACTERISTICS OF ADVANCED REACTORS

This paper is limited to a discussion of the five most widely studied and applied methanogenic retained biomass reactors. Retained biomass reactors for anaerobic waste treatment were first systematically studied with the development of the anaerobic contact process in the fifties. This process was followed by the development of the anaerobic filter. The other processes (upflow anaerobic sludge bed reactors, fluidized or expanded bed reactors and downflow stationary fixed film reactors) were developed in the seventies. Recently, two-phase reactors have been developed based on recent insights into the microbiological and biochemical bases for methanogenesis, but these are not discussed here. More details are available in other recent reviews (Kirsop, 1984; van den Berg and Kennedy, 1983).

Anaerobic Contact Reactor

This was the first retained biomass reactor to be studied and developed systematically (Schroepfer et al, 1955). The principle involved is the same as in the activated sludge process -settling of microbiological floc and other suspended solids and contacting the raw waste with the anaerobic sludge (Fig. 1a). The reactor's performance depends therefore markedly on the efficiency with which the microorganisms and suspended solids settle. An additional factor is the degree to which sludge and raw waste are mixed in the reactor. Good mixing is required yet it is essential that the settling characteristics are not adversely affected. The process is especially suited for wastes with a certain amount of hard-to-digest solids that settle readily or attach themselves readily to settleable solids.

Major limitations of the process are caused by the difficulties of obtaining good settling (Steffen, 1961; van den Berg and Lentz, 1978) and, in large reactors, of providing adequate mixing. Generally, upto 80% of the microorganisms may settle, indicating that the hydraulic retention time cannot be less than one fifth of the minimum mass doubling time. The latter usually is 10 days or more. For many

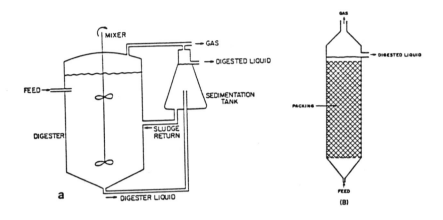

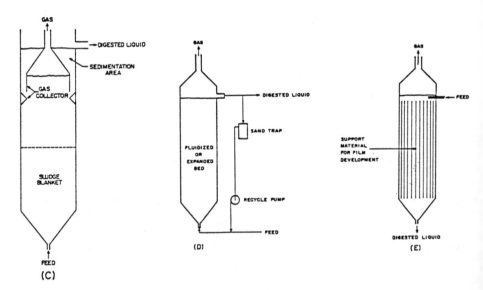

Fig. 1. Sketches of the anaerobic contact (a), anaerobic filter (B), upflow anaerobic sludge bed (C), anaerobic fluidized or expanded bed (D), and downflow stationary fixed film (E) reactors.

wastes, the settling efficiency is less than 80 % and the minimum hydraulic residence times are correspondingly longer.

The anaerobic contact process has been applied commercially in several countries including the U.S.A., Sweden, France and Canada. It is especially useful for wastes containing finely dispersed organic matter, such as starch.

Anaerobic Filter

This reactor, which contains a solid support or packing material, was developed by Young and McCarty (1967). Waste is added at the bottom and flows upward (Fig. 1b). The packing itself serves to separate the gas and to provide quiescent areas for settling of suspended growth. Bacteria are retained mostly in suspended form with a relatively small portion attached to the surface (Dahab and Young, 1982; van den Berg and Lentz, 1979). Since the suspended growth tends to collect near the bottom of the reactor, most of the activity is there. The growth on the surfaces of the packing provides a polishing action (Dahab and Young, 1982; Young and Dahab, 1982). The process is particularly suitable for dilute soluble wastes, soluble wastes which can be made dilute by recirculating effluent or wastes with easily degradable suspended material.

The main limitation of the process is due to accumulation of solids in the packing material which may plug the reactor (Young and Dahab,1982). The solids can be waste suspended solids, material precipitated from the waste (e.g. calcium carbonate) or suspended growth. In addition, hard-to-digest suspended solids that settle readily interfere with the operation of the reactor. In large reactors, an inadequate liquid distribution system may cause channelling and short circuiting.

In spite of the large amount of research done on anaerobic filters, and the great advantages for many wastes, relatively few large, commercial systems appear to have been installed (Witt et al, 1979) . Generally, the potential for plugging and the difficulties of ensuring an adequate flow distribution in the bottom of the reactor appear to have limited application.

Upflow Anaerobic Sludge Bed Reactor

This reactor was developed to avoid the main problem of the anaerobic filter, namely plugging of the packing by suspended growth of bacteria. The packing was reduced to a simple gas collection device that encouraged settling of suspended solids (Fig. 1c) . Since suspended growth in anaerobic reactors usually does not settle well, performance is critically dependent on the development of a readily settling sludge. Lettinga and coworkers were able to develop a granular sludge which settles very well in the reactor (Fig. 2) (de Zeeuw and Lettinga, 1980; Lettinga et al, 1980, 1981, 1982; Hulshoff Pol et al, 1982). As a result, the upflow anaerobic sludge bed (UASB) reactor has operated with high concentrations of microbial biomass, very high loading rates and excellent COD removals, and is particularly suitable for dilute wastes.

Problems with the UASB reactor are usually associated with the development of the granular sludge (Hulshoff Pol et al, 1982, van den Berg et al, 1981). While certain wastes result in a granular sludge quite readily (sugar processing waste and wastes containing mainly volatile acids), other wastes develop this granular sludge very slowly or not at all. Inoculation with a large amount of granular sludge from a well functioning UASB can help. The sludge retains its characteristics often, but not always, when changing from one waste to another.

The process has been applied widely in the Netherlands, Belgium, the U.S.A. and Cuba.

Anaerobic Fluidized and Expanded Bed Reactors

These reactors are similar to suspended growth reactors in that the active biomass is present in the form of a bed of readily settleable aggregates. These aggregates consist of biomass grown on small, inert particles such as fine sand or alumina. A rapid and even flow of liquid is used to keep the particles in suspension (Fig. 1d). The rate of liquid flow and the resulting expansion of the bed (10-25%) determine whether the reactor is called a fluidized or an expanded (10-15% expansion) bed reactor.

The preferred waste substrate for these reactors is soluble or at least the suspended material should be easily degradable (Switzenbaum, 1982).

Reactor performance depends very much on the evenness of the flow of the liquid and as a result the system of liquid distribution is very critical (Jewell, 1982). The capital cost of the flow distribution system and the pumps is high, and also the net energy yield is lower than for other reactors.

The first large scale anaerobic fluidized and expanded bed reactors are under construction in the U. S. A.

Downflow Stationary Fixed Film Reactor

This reactor was also developed to avoid the plugging problems of the anaerobic filter. The reactor contains packing and is operated in the downflow mode (Fig. 1e). Suspended growth and indigestible waste suspended solids are removed with the efffluent. The need for an elaborate distribution system is also eliminated because waste entering at the top of the reactor is readily dispersed by the gas escaping from the packing (Duff and Kennedy, 1983a). The important factor in this reactor is the formation and stability of an active biomass film on the surfaces provided (Murray and van den Berg, 1981; van den Berg and Kennedy, 1980). To avoid accumulation of non-active suspended material in the packing, the architecture of the packing is important.

The downflow stationary fixed film (DSFF) reactor is capable of handling a wide variety of wastes, from reasonably dilute to very concentrated ones (Kennedy and van den Berg, 1982a, 1982b; van den Berg and Kennedy, 1982). Suspended solids, such as those present in some food processing wastes and in manures, are readily accommodated, although their degradation depends on the time they spend in contact with active biomass. The reactor can operate over a wide range of temperatures (Duff and Kennedy, 1983b; Kennedy and van den Berg, 1982c,d).

Loading rates are limited by the amount of active biomass that can be retained in the reactor (Kennedy and van den Berg, 1982b). Effective film thickness is limited by diffusion; hence, the amount of biomass is a function of the surface area available for film formation. This area is limited to less than 100 m^2/m^3

because the channels in the packing have to be large enough to prevent filling up with film. Another limitation is on the use of very dilute waste. To obtain reasonable high loading rates with dilute wastes, the hydraulic retention time has to be short. Since the channels have a minimum dimension, the probability of contact between waste organics and film decrease with decreasing hydraulic retention time.

The process has been applied commercially in Puerto Rico and in Canada.

FUTURE DEVELOPMENTS

The reasons for the use of anaerobic treatment with methane recovery rather than aerobic treatment for waste waters are compelling: net energy production instead of energy consumption, much higher loading rates, simpler operation and reduced capital costs. Since anaerobic methane production is not a complete waste treatment system, and its effluent needs to be treated further, it is necessary to reverse present treatment philosophy: where at all possible, anaerobic treatment should precede, rather than follow aerobic treatment.

Further improvements in efficiency and rates of production, and reduction in costs of operation are possible in the near future. Basic microbiological research in Canada and elsewhere will undoubtedly lead to better control systems, faster rates of reactions and higher conversion efficiencies (for example, by faster hydrolysis of cellulose, proteins and fats, by choosing optimum conditions for what are now largely unknown microorganisms). Work with cocultures (Laube and Martin, 1983) is starting to show promising relationships between different types of bacteria in anaerobic reactors.

It is possible to improve performance of the advanced reactors further, or modify them to become cheaper to install and simpler to operate. The rapidly developing understanding of the microbial and biochemical processes involved, especially those involving ecological relationships, will be of great value in obtaining further improvement. The potential of the sludge bed and expanded bed reactor for the treatment of raw sewage, for example, has not been reached as yet. Combinations of the reactor types discussed are being researched agressively and other methods of cell retention will undoubtedly be invented and developed (Bachmann et al, 1982; Binot et al,1982; Guiot and van den Berg, 1984; Martensson and Frostell, 1982; Oleszkiewicz and Olthoff, 1982) .

Applied research in Canada is aimed at more efficient and cheaper reactor designs, and automatic and less expensive reactor operation. Important for this is a side-by-side comparison of several advanced reactors. This is done by Environment Canada in cooperation with the National Research Council, in a pilot plant (Hall, 1982; Hall et al, 1982). On the other end of the scale (slow rates of methane production), a Canadian firm has been successful commercially in upgrading the old anaerobic lagoon system and is installing methane recovery systems for their "bulk volume reactor" (Landine et al, 1981).

The practical developments so far have provided waste treatment engineers with effective and inexpensive methods of waste treatment which undoubtedly will find increasing use in practise. These developments have also opened up a fascinating field for microbiological studies.

274

Bachmann, A., V.L. Beard, and P.L. McCarty. 1982. Comparison of fixed film
reactors with a modified sludge blanket reactor. Proc. 1st. Int. Conf. Fixed
Film Biol. Processes, 1192-1211.
Binot, R.A., T. Bol, H.T. Naveau, and E.J. Nyns. 1983. Biomethanation by
immobilized fluidized cells. Water Sci. Technol., 15: 103-116.
Dahab, M.F., and J.C. Young. 1982. Retention and distribution of biological
solids in fixed-bed anaerobic filters. Proc. 1st Int. Conf. Fixed Film Biol.
Processes, 1337-1351.
de Zeeuw, W., and G. Lettinga. 1980. Acclimation of digested sewage sludge during
start-up of an upflow anaerobic sludge blanket (UASB) reactor. Proc. 35th Purdue
Indust. Waste Conf., 39-47.
Duff, S.J.B., and K.J. Kennedy. 1983a. Effect of effluent recirculation on start-
up and steady-state operation of the downflow stationary fixed film (DSFF)
reactor. Biotech. Letters, 5(5): 317-320.
Duff, S.J.B., and K.J. Kennedy. 1983b. Effect of hydraulic and organic
overloading on thermophilic downflow stationary fixed film (DSFF) reactor.
Biotech. Letters, 4(12): 815-820.
Guiot, S.R., and L. van den Berg. 1984. Performance and biomass retention of an
upflow anaerobic reactor combining a sludge blanket and a filter. Biotechnol.
Lett., 6(3): 161-164.
Hall, E.R. 1982. Biomass retention and mixing characteristics in fixed film and
suspended growth anaerobic reactors. Proc. IAWPR Seminar Anaerobic Treatment of
Waste Water in Fixed Film Reactors. Copenhagen.
Hall, E.R., M. Jovanovic, and M. Pejic. 1982. Pilot studies of methane production
in fixed film and sludge blanket anaerobic reactors. Proc. 4th Bioenergy R&D
Seminar, Winnipeg, 475-479.
Hulshoff Pol, L.W., W.J. de Zeeuw, C.T.M. Velzeboer, and G. Lettinga. 1982.
Granulation in UASB-reactors. Water Sci. Technol., 15: 291-304.
Hungate, R.E. 1950. The anaerobic mesophilic cellulolytic bacteria. Bacterial.
Reviews, 14: 1-49.
Jewell, W.J. 1982. Anaerobic attached film expanded bed fundamentals. Proc. 1st
Int. Conf. Fixed Film Biol. Processes, 17-42.
Kennedy, K.J., and L. van den Berg. 1982a. Anaerobic digestion of piggery-waste
using a stationary fixed film reactor. Agr. Wastes, 4: 151-158.
Kennedy, K.J., and L. van den Berg. 1982c. Effect of height on the performance of
anaerobic downflow stationary fixed film (DSFF) reactors treating bean blanching
waste. Proc. 37th Purdue Indust. Waste Conf. 71-76.
Kennedy, K.J., and L. van den Berg. 1982d. Stability and performance of anaerobic
fixed film reactors during hydraulic overloading at 10-35°C. Water Res., 16:
1391-1398.
Kennedy, K.J., and L. van den Berg. 1982e. Thermophilic downflow stationary fixed
film reactors for methane production from bean blanching waste. Biotech.
Letters, 4(3): 171-176.
Kirsop, B.M. 1984. Methanogenesis. Crit. Rev. Biotechnol., 1: 109-159.
Landine, R.C., A.A. Cocci, T. Viraraghavan, and G.J. Brown. 1981. Anaerobic
pretreatment of potato processing waste water - a case history. Proc. 36th
Purdue Indust. Waste Conf., 233-240.
Laube, V.M., and S.M. Martin. 1983. Effect of some physical and chemical
parameters on the fermentation of cellulose to methane by a coculture system.
Can. J. Microbiol. 29(11): 1475-1480.
Lettinga, G., W. de Zeeuw, and E. Ouborg. 1981. Anaerobic treatment of wastes
containing methanol and higher alcohols. Water Res., 15: 171-182.

Lettinga, G., S.W. Hobma, L.W. Hulshoff Pol, P. de Jong, W. de Zeeuw, P. Grin, and
R. Roersma. 1982. Design, operation and economy of anaerobic treatment. Water
Sci. Technol., 15: 177-196.
Lettinga, G., S.W. van Velson, W. Hobma, W. de Zeeuw, and A. Klapwyk. 1980. Use
of the upflow sludge blanket (USB) reactor concept for biological waste water
treatment especially for anaerobic treatment. Biotech. & Bioeng., 22: 699-734.
Martensson, L., and B. Frostell. 1982. Anaerobic waste water treatment in a
carrier assisted sludge bed reactor. Water Sci. Technol., 15: 233-246.
Murray, W.D., and L. van den Berg. 1981. Effect of support material on the
development of microbial fixed films converting acetic acid to methane. J. of
Appl. Bact., 51: 257-265.
Oleszkiewicz, J.A., and M. Olthoff. 1982. Anaerobic treatment of food industry
waste waters. Food Tech., 78-82.
Schroepfer, G.J., W.J. Fuller, A.S. Johnson, N.R. Ziemke, and J.J. Anderson. 1955.
The anaerobic contact process as applied to packinghouse wastes. Sewage Ind.
Wastes, 27: 460-486.
Steffen, A.J. 1961. Operating experiences in anaerobic treatment of packinghouse
waste. Proc. 3rd Research Conf., Am. Meat Indust. Foundation, 81-89.
Switzenbaum, M.S. 1982. A comparison of the anaerobic filter and the anaerobic
expanded/fluidized bed processes. Water Sci. Technol., 15: 345-358.
van Brakel. J. 1980. The Ignis Fatuus of biogas. Delft University Press.
van den Berg, L., and K.J. Kennedy. 1980. Support materials for stationary fixed
film reactors for high-rate methanogenic fermentations. Biotech. Letters, 3(4):
165-170.
van den Berg, L., and K.J. Kennedy. 1982b. Effect of substrate composition on
methane production rates of downflow stationary fixed film reactors. Proc. IGT
Symp. Energy from Biomass and Wastes, 401-424.
van den Berg, L., and K.J. Kennedy. 1983. Comparison of advanced reactors. Proc.
3rd. Intern. Symp. Anaerobic Digestion, Boston, Mass., 71-89.
van den Berg, L., K.J. Kennedy, and M.F. Hamoda. 1981a. Effect of type of waste
on performance of anaerobic fixed film and upflow sludge bed reactors. Proc. of
36th Purdue Indust. Waste Conf., 686-692.
van den Berg, L., and C.P. Lentz. 1978. Factors affecting sedimentation in the
anaerobic contact fermentation using food processing wastes. Proc. 33rd Purdue
Indust. Waste Conf., 185-193.
van den Berg, L., and C.P. Lentz. 1979. Comparison between up- and downflow
anaerobic fixed film reactors of varying surface-to-volume ratios for the
treatment of bean blanching waste. Proc. 34th Purdue Indust. Waste Conf., 319-
325.
Witt, E.R., W.J. Humphrey, and T.E. Roberts. 1979. Full-scale anaerobic filter
treats high strength wastes. Proc. 34th Purdue Indust. Waste Conf., 229-234.
Young, J.C., and M.F. Dahab. 1982. Effect of media design on the performance of
fixed bed anaerobic reactors. Water Sci. Technol., 15: 369-384.
Young, J.C., and P.L. McCarty. 1967. The anaerobic filter for waste treatment.
Proc. 22nd Purdue Indust. Waste Conf., 559-574.

SCALE-UP OF THE BIO-HOL PROCESS FOR THE CONVERSION OF

BIOMASS TO ETHANOL

G.R. Lawford, R. Charley, R. Edamura, J. Fein, K. Hopkins, D. Potts,
B. Zawadzki, Weston Research Centre, 1047 Yonge Street, Toronto,
Ontario, M4W 2L3, and H. Lawford, Dept. of Biochemistry, University
of Toronto, Toronto, Ontario, M5S 1A8.

ABSTRACT

A state-of-the-art pilot plant with a capacity for the continuous
conversion of 1 oven dried ton of wood per day has been designed and
equipped for process optimization and data acquisition in the
demonstration of the Bio-hol process (European patent 0047641).
Previous studies reported by ourselves and others have shown that the
process microorganism, Zymomonas mobilis, has considerable advantages
(higher product yield, specific and volumetric productivity) than
comparable yeast-based, laboratory-scale continuous processes. These
advantages have been confirmed at pilot-scale, using several process
configurations. A continuous process has also been developed at the
pilot plant for producing inexpensive sugar hydrolysates from wood
and agricultural residues, based on a custom designed extruder. Data
will be presented on the fermentation performance of both
Z. mobilis and Saccharomyces cerevisiae on this feedstock, and on
current efforts for increasing productivity through feedstock
refining and strain optimization.

INTRODUCTION

In traditional fuel alcohol fermentation processes using a corn-
starch hydrolysate or molasses as a feedstock, the high cost of raw
materials is the most significant factor in the overall process
economics (Flannery and Steinschneider, 1983). Since the substrate
cost in these processes represents 50-60% of the total manufacturing
expenses, increased substrate conversion efficiency and/or the use of
potentially less expensive raw materials, eg. hydrolysates of lignocellulosic wastes, would significantly reduce overall production
costs. Further reductions could also be achieved by increased
productivity and also by decreased product recovery costs as a result
of an elevated alcohol concentration in the fermentation broth.

Although the conversion of lignocellulosics into fermentable sugars has attracted considerable research interest (Bungay, 1984), development of a controlled, reproducible system for production of inexpensive hydrolysate sugar streams has remained elusive. To be suitable for eventual commercial-scale operation the conversion process should ideally be capable of both pretreatment and single-pass hydrolysis (to maximize downstream flexibility), simple in design and operation (to ensure reliability and safety), energy efficient and high productivity (eg. continuous).

Available conversion technologies were reviewed and evaluated during development of the Bio-hol process design. Extrusion technology appeared very attractive due to the low energy demand, simplicity and flexibility of design, facility of scale-up and the inherent continuous nature of the process. Following some promising preliminary studies at Wenger Mfg., Sabetha, Kansas (unpublished results) sufficient data were generated to permit construction of a custom-built pre-prototype unit for design-refinement and feedstock-conversion yield studies at the Bio-hol pilot plant.

Throughout the preliminary lignocellulose conversion trials at Wenger, initial promising laboratory fermentation results reported using the bacterium Zymomonas mobilis (Lavers et al, 1981) were being developed into a series of advanced process designs to fully exploit the well-documented advantages of this bacterium (Rogers et al, 1982; Lawford et al, 1983). The reported increased substrate conversion efficiency and higher specific alcohol formation rate compared to the traditionally employed yeast strains, and the ability of Z. mobilis to continue high-rate alcohol production in growth-restricting conditions, were exploited in single-stage continuous fermentations using free and flocculated cell strains, operating at higher efficiency than comparable yeast processes, and in a two-stage continuous process producing an effluent containing >100g/L alcohol (Lawford et al, 1983).

While preliminary evaluation of extrusion technology continued, a flexible, state-of-the-art continuous culture fermentation pilot plant was assembled at the Bio-hol facility to allow evaluation of the laboratory-demonstrated advanced Zymomonas fermentation process designs at pilot scale.

METHODS

i) Extrusion

Extrusion studies were carried ot using a custom-modified single-screw extruder (Model SX80, Wenger MFG., Sabetha, Kansas). Initial work concentrated on design modifications to reduce wear and corrosion and to provide stable, safe operation. Operating pressures and temperatures were achieved as a sole result of mechanical energy generated by rotation of the extruder screw and were readily controllable by operators during both pretreatment and acid hydrolysis runs.

By varying operating conditions a range of pretreated sawdust (pine and poplar) products were produced, ranging from fibrous (mild pretreatment) to amorphous (harsh pretreatment), while controlled injection of sulphuric acid permitted the production of acid hydrolysed wood streams. Acid hydrolysed products were similarly produced using canola stalks, corn stover, wheat straw and soya stalks as feedstocks.

ii) Pilot-scale Fermentation

All pilot-scale fermentation trials were carried out using custom-designed equipment (New Brunswick Scientific Co., Edison, New Jersey) in the following continuous culture process designs:

a) single-stage free cell continuous culture (operating volume 400 litres);

b) single-stage continuous culture with flocculent strain and cell recycle by external settling (total system operating volume 100 litres);

c) two-stage continuous culture (stage I operating volume 100 litres; stage II operating volume 400 litres);

d) two-stage continuous culture with cell recycle on the second stage (stage I operating volume 400 litres; stage II total system volume 215 litres; cell recycle by ultrafiltration using Iopor 8C system, Dorr-Oliver, Stamford, Conn.).

In all fermentation studies, single- or first-stage fermentations were fed with a sterile medium containing starch hydrolysate (Staleydex 333) as a carbon source. Second stage fermentations were fed with an unsterilized concentrated starch hydrolysate solution.

iii) Hydrolysate Fermentability Trials

The suitability of hydrolysate streams as fermentation feedstocks was assessed by minimum inhibitory concentration (MIC) tests using the broth dilution technique (Anderson, 1970). Dilution was carried out using medium containing 50 g/l Staleydex and 5 g/l technical yeast extract (TYE), while undiluted samples were supplemented with 5 g/l TYE.

Growth was determined by macroscopic and microscopic evaluation after 1 and 3 days.

iv) Analytical Procedures

Moisture content was determined by weight difference after drying to constant weight in a vacuum oven at 60°C.

Hydrolysate sugar concentration was determined by high resolution capillary gas chromatography.

Cellulose content was determined by hydrolysis to soluble sugars with sulphuric acid and measurement of sugar concentrations as above.

"Active" or "pristine" lignin concentration was determined as the fraction of residue soluble in acetone following aqueous extraction.

Ethanol and starch hydrolysate glucose were determined by high performance liquid chromatography.

RESULTS AND DISCUSSION

i) Sawdust Pretreatment

Since the major thrust of the Bio-hol lignocellulose conversion project was to develop a single-stage hydrolysis system, the majority of the effort on the extrusion system was directed towards operation with acid injection for hydrolysis. However, the effect of extruder operating conditions without acid injection was correlated with the digestibility of products obtained by cellulase and hemi-cellulase enzyme hydrolysis (Table 1), providing preliminary data for use of the extruder for substrate pretreatment.

Table 1: Enzyme digestibility of sawdust pretreated under various extruder operating conditions

Feedstock	Average Pressure[1] (psig)	Temperature[1] (°F)	Enzyme[2] Digestibility (%)
Poplar A	700	360	37
Poplar B	1000	420	49
Poplar C	1750	440	54
Pine A	1500	430	29
Pine B	2000	440	36

Notes: 1- Measured at outlet head.
2- Calculated on dry product basis.

Additionally, the sugar composition of the filtrates from enzyme digestions were analysed (Table 2), while acetone extraction was used as an estimate of the concentration of "active" or "pristine" lignin in the pretreated material. Typically, using poplar sawdust as a feedstock, an enzyme digestibility of 50% could be routinely achieved, giving an enzyme soluble fraction containing 42% glucose, 4% xylose and 17% "active" lignin on dry product basis (Lawford et al, 1984).

Table 2: Sugar composition of enzyme soluble fractions of pretreated poplar sawdust samples.

Component	Sugar Composition (%)[1]		
	Poplar A	Poplar B	Poplar C
Glucose	15	21	34
Xylose	16	17	7.2
Mannose	1.4	1.6	1.1
Arabinose	0.3	0.2	--
Galactose	0.1	--	--

Note: 1- Calculated on a dry product basis.

ii) Single-stage Hydrolysis

By injection of sulphuric acid into the terminal portion of the extrusion system, single-stage hydrolysis of sawdust into sugar streams was achieved (Table 3). By autoclaving samples prior to analysis, the concentration of sugar monomers was significantly increased (figures in parenthesis, Table 3), presumably due to decomposition of oligomeric species. This suggests that a significant increase in yield of sugar monomers could be achieved by use of a mild secondary hydrolysis step.

Table 3: Composition of hydrolysed sawdust samples.

Component	Composition on dry product basis (%)	
	Pine	Poplar
Aqueous solubles	44	43
- Glucose	21 (28)	24 (31)
- Xylose	1.0 (1.1)	5.8 (7.7)
- Mannose	3.4 (4.7)	1.1 (2.1)
- Galactose	1.2 (1.5)	0.1 (0.1)
- Arabinose	0.4 (0.5)	--
Residual cellulose	12	24
Residual lignin	44	32

Note: Figures in parenthesis are concentrations measured following autoclaving at 15 psig for 1h..

Although the extrusion system used in these trials was substantially
refined to permit stable continuous operation as a single-stage
hydrolysis system for sawdust feedstocks, preliminary hydrolysis
trials using agricultural lignocellulosic residues were successful
(Table 4). It is possible that significant increases in glucose
yield could be achieved by use of a purpose-built extrusion system
and further studies to optimize extrusion conditions.

Table 4: Glucose yields from hydrolysis of agricultural residues

Feedstock	Cellulose Content (% Raw Material)	Maximum Glucose Yield (after autoclaving)
Soya Stalks	38.0	18
Wheat Straw	37.9	18
Canola Stalks	25.7	6.6
Corn Stover	33.7	13

iii) Pilot-scale Fermentation

The results presented in Table 5 show that the laboratory-demonstrated
advantages of Zymomonas fermentations have been successfully
reproduced at pilot scale. Throughout operations at the pilot plant
during the period June 1983 - June 1984, the high substrate conversion
factor quoted for Zymomonas has been achieved. This would represent
an overall saving of up to 2% total production costs, compared to
traditional yeast fermentation processes, in commercial production.

Additionally, the flocculent cell process design has produced a
significant increase in process productivity compared to all the
processes shown in Table 5. This would translate into a significant
reduction in the size of plant required for a given production volume
and, therefore, in capital investment and production costs. This
demonstrated benefit may be further improved by ongoing cell
settlement optimization trials.

Operation of the two-stage Bio-hol process design has permitted high
(>100 g/L) product alcohol concentrations to be achieved, and use of
cell recycle by ultrafiltration on the second stage has substantially
increased process productivity. Operation of the two-stage design,
with our without cell recycle, at commercial scale would reduce
production costs due to decreased cost of product recovery (by
distillation).

Table 5: Comparison of pilot-scale operation of advanced continuous alcohol fermentation processes.

Process		Product Alcohol Concentration (g/l)	Process Yield (% Theoretical)	Productivity (g/l.h)
Bio-hol Develop-ments	Single-stage (free cell)	53-60	93-97	9-13
	Single-stage (flocculated cell)	58-60	93-97	24-25
	Two-stage (without cell recycle)	96-105	88-98	4-5
	Two-stage (with cell recycle)	95	97	14
Pfiefer and Langen[1] (Zymomonas)		56	92	4.5
Alcon[2] (Yeast)	Molasses	57	88-91	8
	Cane Juice	65	91	9
Hoechst/Uhde[3] (Yeast)		80	90	16
National Distillers[4] (Yeast)		90-100	88	Not Available

Notes: 1 - Single stage continuous culture (stired tank reactor); free cell Zymomonas system (Bringer et al, 1984)
2 - Single-stage continuous culture (stirred tank reactor) with yeast recycle by external sedimentation (Alcon, 1983)
3 - Single-stage continuous culture (air lift tower reactor) with yeast recycle by external sedimentation (Faust et al, 1983)
4 - Two-stage continuous culture (stirred tank reactors) with yeast recycle by continuous centrifugation (Muller and Miller, 1983)

All yield data calculated on input sugar concentration.

283

) Fermentability of Hydrolysate Streams

The toxicity of unrefined acid wood hydrolysates to the growth and
fermentative activity of yeast strains has been reported (Leonard and
Hajny, 1945; Fein et al, 1984). During the Bio-hol project
considerable effort has been directed to developing methods for the
removal of inhibitory compounds from hydrolysates of both sawdust and
agricultural residues.

While the raw sawdust hydrolysate produced by the Bio-hol extrusion
process proved inhibitory to both Z. mobilis and the yeast
S. cerevisiae at low concentration, significant increases in
tolerance were achieved using proprietary methods developed for
removal of inhibitors (Table 6).

Currently, research is continuing to develop alternate procedures for
inhibitor removal and to produce process organism strains more
tolerant to inhibitory compounds.

Table 6:   Growth of process organisms on wood hydrolysate streams

Hydrolysate Stream	MIC*(% w/v hydrolysate glucose)	
	Z. mobilis	S. cerevisiae
Raw hydrolysate	0.8	1.5
Partially refined hydrolysate	3.4	6.8
Refined hydrolysate	16.9	>16.9

*Minimum inhibitory concentration

Preliminary studies using hydrolysates obtained from agricultural
residues have indicated that the raw hydrolysate streams might be
less inhibitory than raw sawdust hydrolysates (Table 7). This would
be important, since the production of high fermentable glucose
concentrations might, therefore, be economically more favourable
using agricultural residues as feedstocks.

284

Table 7:   Growth of process organisms on raw biomass hydrolysates.

Hydrolysate Stream	MIC (% w/v hydrolysate glucose)	
	Z. mobilis	S. cerevisiae
Poplar	0.8	1.5
Canola Stalk	1.2	>1.2
Corn Stover	0.5	>0.5
Soya Stalk	2.8	>2.8
Wheat Straw	3.3	>3.3

Conclusions:

1. A controlled, energy efficient continuous process has been developed for high yield conversion of lignocellulosics to sugar streams by single-stage hydrolysis.

2. Overall yield of the hydrolysis process could be significantly increased by a mild secondary hydrolysis.

3. Continuous substrate pretreatment has also been demonstrated.

4. High productivity, high yield Zymomonas fermentations have been demonstrated at pilot-scale.

5. High product alcohol concentrations have been achieved by use of two-stage fermentation process designs at pilot scale.

6. Significant improvements in hydrolysate fermentability have been achieved by inhibitor removal.

Acknowledgements:

The Bio-hol Developments venture has had financial support through the Ontario Ministry of Energy, and Energy, Mines and Resources, Canada (CREDA).

285

REFERENCES

Alcon, (1983) "Tropical demonstration of the Alcon process", Alcon
Biotechnology Ltd. technical leaflet.

Anderson, T.G. (1970) In: Manual of Clinical Microbiology, J.E.
Blair, E.H. Lennette and J.P. Triant (eds.), Amer. Soc.
Microbiol., (Baltimore) pp. 299-310.

Bringer, S., H. Sahm and W. Swyzen (1984) "Ethanol production by
Zymomonas mobilis and its application on an industrial scale",
presented at 6th Symposium on Biotechnology in Energy
Production and Conservation, Gatlinburg, Tennessee, USA.

Bungay, H.R. (1984) "A timetable for commercializing biomass
refining", presented at 3rd European Congress on Biotechnology,
Munich, W. Germany.

Faust, U., P. Prave and M. Shlingmann (1983) Process Biochem. 18 (3),
31-37.

Fein, J.E., S.R. Tallim and G.R. Lawford (1984) Can. J. Microbiol.
30, 682-690.

Flannery, R.J. and A. Steinschneider (1983) Biotechnology 1,
773-776.

Lavers, B.H., P. Pang, C.R. MacKenzie, G.R. Lawford, J. Pik and H.G.
Lawford (1981) In: Advances of Biotechnology. Vol. 2, eds. M.
Moo-Young and C.W. Robinson (eds), Pergamin Press Canada Ltd.
pp. 195-200.

Lawford, G.R., B.H. Lavers, D. Good, R. Charley, J. Fein and H.G.
Lawford (1983) In: Proceedings on the Royal Society of Canada
International Symposium on Ethanol from Biomass. H.E. Duckworth
and E.A. Thompson (eds), Royal Society of Canada (Ottawa).

Lawford, G.R., R. Charley, R. Edamura, J. Fein, K. Hopkins, D. Potts,
B. Zawadzki and H. Lawford (1984) "Biomass to ethanol by the
Bio-hol process", presented at 5th Bioenergy Symposium, Ottawa,
Ontario, Canada.

Leonard, R.H. and G.H. Hajny (1945) Ind. Eng. Chem. 37, 390-395.

Muller, W.C. and F.D. Miller (1983) U.S. Patent No, 4,385,118.

Rogers, P.L., and K.J. Lee, M.L. Skotnicki and D.E. Tribe (1982) In:
Advances in Biochemical Engineering, Vol. 2., A. Feichter (ed)
Springer-Verlag, (New York), pp. 37-84.

COMMERCIAL UPDATE ON THE STAKE PROCESS

J. D. Taylor

Stake Technology Limited
208 Wyecroft Road
Oakville, Ontario
L6K 3T8

ABSTRACT

The history of Stake Technology, is summarized beginning in 1973, as the first commercial company using auto-catalysed hydrolysis as the principal technology for the conversion of waste biomass into organic chemicals and ruminant feeds. A brief outline of commercial installations, large scale research trials in animal feed and chemical production, sales effort, international agreements and a near-term outlook for the use of the continuous Stake Biomass Process (SBP) will also be included. SBP is a Stake development specifically created for the isolation of soluble hemicellulose, lignin and cellulose, with post-treatment for higher value-added products.

INTRODUCTION

Stake Technology Ltd. was founded in 1973 to develop and market a process for the conversion of waste biomass into ruminant animal feed and organic chemicals. The company set out to use commercially available equipment to perform the steam treatment required to upgrade the digestibilty of these ligno-cellulosic materials. The company experimented with several different types of commercially available equipment, however none of this equipment was capable of operating on an economic basis. The company subsequently developed its own proprietary equipment capable of operating in a continuous mode.

The company has conducted over 60 animal feeding trials in conjunction with univerities and other research institutions. These studies have shown the Stake process to produce high energy fibre cattle feeds from a wide range of raw materials including hardwoods, straws, cornstover and sugarcane bagasse. The company has demonstrated its cattle feed process in commercial facilities located in the U.S.A.

287

In 1976, the company began its research efforts into other applications of Stake treated materials. The company's developmental work in this area resulted in the Stake Biomass Process (SBP). SBP is a process which effectively separates lignocellulosic materials into their three main components, hemi-cellulose, lignin and cellulose (see Figure 1). The cellulose fraction generated in SBP can be hydrolyzed (by acid or enzymes) for fermentation to ethanol. SBP has been demonstrated in trials conducted at tonnage scale in the fractionation of a range of raw materials.

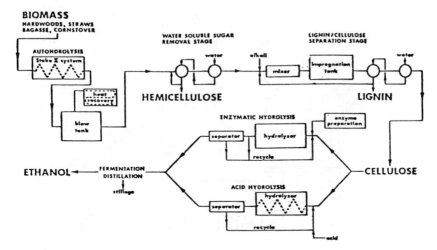

Figure 1:
Stake Biomass Conversion Process

The purpose of this presentation is to highlight the commercial side of Stake's development - where we've been and where we are going. I will first describe the evolution of the Stake II system from Stake's original investigations of batch systems to its most modern proprietary continuous equipment. I will then describe Stake's approach to the marketing of its technology from originally being an equipment supplier to its present position to being a licensor of technology. I will then speak briefly on the installations that Stake has built or is currently in the process of building and finally I will present a summary of the major commercial prospects facing Stake in the near term.

EQUIPMENT

Stake began its development work using a batch reactor system. The company subsequently concluded that batch processing placed limitations on the process in an economic sense as well as a technical sense. The company realized that processing on a continuous basis would not only reduce operating costs by lowering energy consumption but would also provide better control over reaction conditions to maximize product quality and consistency. The company's first attempt at development of a continuous system employed a commercially available extruder feeder.

288

The company achieved a limited degree of success with this system, however the system was not capable of operating continuously on a reliable basis and wear rates were unacceptable. The company subsequently developed the patented co-ax feeder, which has become the heart of the Stake II system. (See Figure 2)

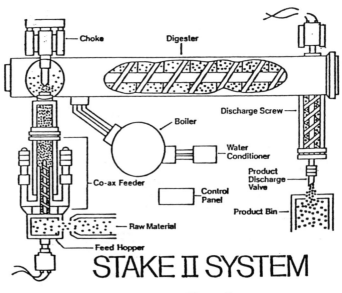

Figure 2

The first co-ax feeder manufactured by Stake was a hydraulically driven 6" diameter feeder installed as part of a mobile system constructed to demonstrate the company's cattle feed technology at various research institutions located across the continent. This first continuous Stake II system had a capacity of one-half ton of wood chips per hour. The company subsequently scaled up its equipment by designing a mechanically driven 8" feeder. This system was capable of processing 4 tons of wood chips per hour and has been the basis of Stake's commercial facilities established to date in the U.S.A..

The company has completed design of a 10" diameter feeder capable of processing 10 tons of wood chips per hour. Construction of the first 10" prototype feeder will be completed in 4 months and this feeder will become the heart of Stake systems marketed in the future.

MARKETING

Stake originally set out to sell biomass conversion equipment. The company soon discovered that the task at hand was significantly greater than had been expected. Cattlefeed products from biomass were not familiar to the farmer and it became obvious that an increased marketing effort was required. It was not realistic to expect a small saw mill operator to be able to market a novel cattlefeed product. Stake subsequently adopted a joint venture approach to the marketing of its technology. The basic philosophy was that Stake would enter into agreements with raw material suppliers and retain an equity position in the joint venture. Stake in this approach would be in a position to more fully support the marketing efforts of such an operation. Stake establised 2 joint venture facilities in the U.S.. The joint venture philosophy however, limited the company's growth. Distant markets could not be effectively served by the joint venture approach and the approach required a significant investment on behalf of the company.

Stake subsequently embarked on its current licensing of technology approach to marketing. Approximately one year ago, the French engineering group "Technip" acquired the right to market Stake's cattlefeed process on a world wide basis and the Stake Biomass Process world wide outside of the U.S.A..

Technip is a large engineering firm which operates on a world wide basis with particuliar experience with projects in the third world and the Comecon countries. The association with Technip has opened up markets which in the past have been difficult for Stake to enter.

The rights ro marketing of SBP in the U.S.A. are held by the engineering firm of Vulcan Cincinnatti Inc.. Vulcan has been active in biomass utilization for decades and has extensive knowledge of alcohol production. The Vulcan/Stake combination combines the expertise necessary for large scale conversion of lignocellulosic biomass into ethanol.

INSTALLATIONS

Stake has built 2 joint venture cattlefeed demonstration plants in the U.S.A.. The first was a joint venture with a sugar company and the plant manufactured animal feed in the silage form using sugarcane bagasse as a raw material. The plant was located on a 6,000 head feed lot which purchased all of the plants output. Due to an overall decline in the beef industry in south Florida, the feedlot subsequently closed down.

In order to market the product it then became necessary to ship it significant distances for utilization by the dairy industry. The extra handling and transportation costs dictated that the product be in pelletized form. Investing in drying and pelletizing facilities would have represented a significant capital investment which was not considered to be an attractive investment due to the depressed state of the Florida cattle feed market. The plant was subsequently used as a demonstration facility for two years and we are now in the process of closing down the facility.

In 1980, the company established a joint venture in Minnesota to produce a silage cattle feedproduct from the waste of a pallet manufacturer. The product sold commercially for a short period of time however drastic declines in feed prices made the plant uneconomic due to competition from conventional feeds. The Minnesota facility is currently being used as Stake's primary commercial demonstration facility.

Stake is currently in the process of building two additional facilties. A Stake unit in Soustons in France will be part of the French Gasohol Plan. The Stake system will be the pretreatment used to treat wood, cornstover and straw for the production of acetone-butanol.

Stake is also currently in the process of installing a research facility at the University of Sherbrooke. The research team will be headed up by Dr. Esteban Chornet. Dr. Chornet and his group are one of the major groups researching biomass conversion in Canada. We feel that the relationship with the University will be highly beneficial to both parties.

CURRENT PROSPECTS

On the cattlefeed side, Stake recently completed animal feeding trials in China and Russia. The trials confirmed that the products can be effectively used in the rations employed in these countries and negotiations toward establishing production facilities are currently underway through Technip.

Detailed negotiations are also underway for the construction of a Stake facility to be located in the Middle East to convert sugarcane bagasse into furfural and a pelletized cattle feed product.

Through Vulcan Cincinatti, Stake is pursuing two major SBP projects in the U.S.A.. One project is located in upstate New York and is designed to convert underutilized hardwoods such as red maple into hemi-cellulose, lignin and cellulose which will be sold to third parties for conversion into higher value added products.
In conjunction with the New York State Energy Research and Development Agency, a feasibilty study is now nearing completion. A decision will be made in the near future on whether or not to proceed with the construction of the New York facility.

The second major project that Stake is involved in in North America is a project in Florida. The facilty will utilize laurel oak which is a scrub tree in Florida and will produce a hemi-cellulose concentrate and lignin which will be sold to third parties. The cellulose will be hydrolized enzymatically to glucose and will be fermented to produce fuel grade ethanol. Certain tax incentives relative to alcohol fuel production in Florida are favorable and make the economics of this project look extremely promising.

Stake is also actively pursuing the application of its technology to paper making and recently completed some very successful trials. The company is exploring the possibility of installing a system in the Province of Quebec in 1985 to demonstrate its pulping technology.

Stake is continuing to investigate other applications of its feeder technology in such areas as dewatering and pressurized gasification systems.

291

SUMMARY

In summary, I would like to say that the future for Stake Technology
looks extremely bright. The company is now in a position to develop
its cattlefeed process in the areas of the world where it is most
needed. As markets for biomass derived lignins and hemicellulose are
rapidly being developed, Stake's biomass conversion process represents
a major growth opportunity for the company. This combined with the new
applications for the company's technology will make for an exciting
future at Stake.